IMPROVING THE REGULATION AND MANAGEMENT OF LOW-ACTIVITY RADIOACTIVE WASTES

Committee on Improving Practices for Regulating and Managing Low-Activity Radioactive Waste

Nuclear and Radiation Studies Board

Division on Earth and Life Studies

NATIONAL RESEARCH COUNCIL
OF THE NATIONAL ACADEMIES

THE NATIONAL ACADEMIES PRESS
Washington, D.C.
www.nap.edu

THE NATIONAL ACADEMIES PRESS, 500 Fifth Street, N.W., Washington, DC 20001

NOTICE: The project that is the subject of this report was approved by the Governing Board of the National Research Council, whose members are drawn from the councils of the National Academy of Sciences, the National Academy of Engineering, and the Institute of Medicine. The members of the committee responsible for the report were chosen for their special competences and with regard for appropriate balance.

The 10 organizations that provided financial support for this report are recognized in the Preface.

International Standard Book Number 0-309-10142-5 (Book)
International Standard Book Number 0-309-65838-1 (PDF)
Library of Congress Control Number 2006922091

Cover: Typical low-activity waste, courtesy of the Nuclear Energy Institute.
Waste shipping containers, courtesy of MHF Logistical Solutions.
Uranium-contaminated areas, courtesy of the Department of Energy.

Additional copies of this report are available from the National Academies Press, 500 Fifth Street, NW, Lockbox 285, Washington, DC 20055; (800)624-6242 or (202)334-3313 (in the Washington metropolitan area); Internet, http://www.nap.edu.

TD
898.118
.N363
2006

THE NATIONAL ACADEMIES
Advisers to the Nation on Science, Engineering, and Medicine

The **National Academy of Sciences** is a private, nonprofit, self-perpetuating society of distinguished scholars engaged in scientific and engineering research, dedicated to the furtherance of science and technology and to their use for the general welfare. Upon the authority of the charter granted to it by the Congress in 1863, the Academy has a mandate that requires it to advise the federal government on scientific and technical matters. Dr. Ralph J. Cicerone is president of the National Academy of Sciences.

The **National Academy of Engineering** was established in 1964, under the charter of the National Academy of Sciences, as a parallel organization of outstanding engineers. It is autonomous in its administration and in the selection of its members, sharing with the National Academy of Sciences the responsibility for advising the federal government. The National Academy of Engineering also sponsors engineering programs aimed at meeting national needs, encourages education and research, and recognizes the superior achievements of engineers. Dr. Wm. A. Wulf is president of the National Academy of Engineering.

The **Institute of Medicine** was established in 1970 by the National Academy of Sciences to secure the services of eminent members of appropriate professions in the examination of policy matters pertaining to the health of the public. The Institute acts under the responsibility given to the National Academy of Sciences by its congressional charter to be an adviser to the federal government and, upon its own initiative, to identify issues of medical care, research, and education. Dr. Harvey V. Fineberg is president of the Institute of Medicine.

The **National Research Council** was organized by the National Academy of Sciences in 1916 to associate the broad community of science and technology with the Academy's purposes of furthering knowledge and advising the federal government. Functioning in accordance with general policies determined by the Academy, the Council has become the principal operating agency of both the National Academy of Sciences and the National Academy of Engineering in providing services to the government, the public, and the scientific and engineering communities. The Council is administered jointly by both Academies and the Institute of Medicine. Dr. Ralph J. Cicerone and Dr. Wm. A. Wulf are chair and vice chair, respectively, of the National Research Council.

www.national-academies.org

COMMITTEE ON IMPROVING PRACTICES FOR REGULATING AND MANAGING LOW-ACTIVITY RADIOACTIVE WASTE

DAVID H. LEROY, *Chair*, Leroy Law Offices, Boise, Idaho
MICHAEL T. RYAN, *Vice Chair*, Charleston Southern University, South Carolina
EDWARD L. ALBENESIUS, Westinghouse Savannah River Company (retired), Aiken, South Carolina
WM. HOWARD ARNOLD, Westinghouse Electric (retired), Coronado, California
FRANÇOIS BESNUS, Institute de Radioprotection et de Sûreté Nucléaire, Paris, France
PERRY H. CHARLEY, Diné College-Shiprock Campus, New Mexico
GAIL CHARNLEY, Health Risk Strategies, Washington, DC
SHARON M. FRIEDMAN, Lehigh University, Bethlehem, Pennsylvania
MAURICE C. FUERSTENAU, Mackay School of Mines, University of Nevada, Reno
JAMES HAMILTON, Duke University, Durham, North Carolina
ANN RAPPAPORT, Tufts University, Medford, Massachusetts
D. KIP SOLOMON, University of Utah, Salt Lake City
KIMBERLY W. THOMAS, Los Alamos National Laboratory, New Mexico

Liaison
ROBERT M. BERNERO, U.S. Nuclear Regulatory Commission (retired), Gaithersburg, Maryland

Staff
JOHN R. WILEY, Study Director
TONI GREENLEAF, Financial and Administrative Associate
DARLA J. THOMPSON, Research Associate
MARILI ULLOA, Senior Program Assistant
LAURA D. LLANOS, Senior Program Assistant
JAMES YATES, JR., Office Assistant

v

List of Reviewers

This report has been reviewed in draft form by individuals chosen for their diverse perspectives and technical expertise, in accordance with procedures approved by the National Research Council's (NRC's) Report Review Committee. The purpose of this independent review is to provide candid and critical comments that will assist the institution in making its published report as sound as possible and to ensure that the report meets institutional standards for objectivity, evidence, and responsiveness to the study charge. The content of the review comments and draft manuscript remain confidential to protect the integrity of the deliberative process. We wish to thank the following individuals for their review of this report:

David Adelman, University of Arizona, Tucson
Jan Beyea, Consulting in the Public Interest, Lambertville, NJ
Robert J. Budnitz, Lawrence Livermore National Laboratory, CA
Michael Corradini, University of Wisconsin, Madison
Sharon Dunwoody, University of Wisconsin, Madison
Gordon Linsley, International Atomic Energy Agency (retired), Oxon, England
Michael McWilliams, Stanford University, CA
Richard Meserve, Carnegie Institution, Washington, DC
Dianne Nielson, Department of Environmental Quality, Salt Lake City, UT
Allan Richardson, Environmental Protection Agency (retired), Bethesda, MD
Atsuyuki Suzuki, Nuclear Safety Commission, Tokyo, Japan

Although the reviewers listed above have provided many constructive comments and suggestions, they were not asked to endorse the conclusions or recommendations, nor did they see the final draft of the report before its release. The review of this report was overseen by John F. Ahearne, Sigma Xi and Duke University, Research Triangle Park, NC. Appointed by the NRC, he was responsible for making certain that an independent examination of this report was carried out in accordance with NRC procedures and that all review comments were carefully considered. Responsibility for the final content of this report rests entirely with the authoring committee and the NRC.

Preface

S tudies by the National Academies provide scientific and technical
advice to assist public decision makers. Studies are typically conducted at the request of a government agency, which funds the
study. This study, however, was self-initiated by the National Academies'
Nuclear and Radiation Studies Board (NRSB). Looking back over 60 years
since the widespread use of nuclear energy began, Board members recognized that statutes, regulations, and commercial practices that deal with
low-activity radioactive wastes—which comprise the largest volume of
radioactive wastes in the United States—have evolved as an inconsistent
patchwork. Low-activity wastes range from medical and laboratory
wastes, to industrial-scale equipment and process residues, to rubble and
contaminated soils from nuclear facility decommissioning and cleanup,
and to mining and mineral extraction wastes. Clearly this wide variety of
wastes touches on many sectors of the economy.

Low-activity wastes are regulated primarily by their origins—the
nature of the industry that produced them—rather than the actual radiological hazards they present. Wastes from some origins are tightly controlled, resulting in limited and relatively expensive management and
disposal options; while other wastes that present equal or greater risks
are less closely controlled.

Once initiated by the NRSB, this study received a great deal of interest
from agencies responsible for the regulation and disposition of low-
activity wastes as well as from public stakeholders. The committee grate-
fully acknowledges the financial support of the following 10 federal, state,
and foreign organizations, which made this study possible:

- Army Corps of Engineers
- California Environmental Protection Agency
- Department of Defense Executive Agent for Low-Level Radioactive Waste
- Department of Energy
- Environmental Protection Agency
- The Institute of Applied Energy—Japan
- Institute de Radioprotection et de Sûrété Nucléaire—France
- Midwest Interstate Low-Level Radioactive Waste Compact
- Nuclear Regulatory Commission
- Southeast Compact Commission

The committee benefited greatly from the diversity of perspectives, concerns, and new ideas brought to our attention by our sponsors. Congressional staff, industry representatives, and members of the public also provided valuable insights. Presentations to the committee (see Appendix C) generally cited needs and opportunities to improve the current system of regulations and management practices, but differed in what specific changes were needed or their urgency. Presenters also cautioned the committee that its advice should be practical and implementable in the context of existing legislation, regulation, and commercial infrastructure.

The first half of this study culminated in an interim report that provided an overview of the current system and identified areas for improvement.[1] In the second half of the study, which led to this final report, the committee developed the concept of a "risk-informed" framework that would provide rationale and structure for significant improvements in the system. By focusing on the risk presented by given wastes, rather than their origin, and requiring consistent measures to control these risks, the framework would further enhance safety, improve efficiency, and promote cooperation among all stakeholders.

While noting current initiatives in the United States and internationally that are sound examples of risk-informed practices, the committee did not suggest specific changes in current legislation, regulations, or commercial practices. Rather it is the committee's position that specific changes are matters of public policy to be developed through the risk-informed decision-making structure set forth in this report.

The committee especially recognizes the efforts by the members and staff of the NRSB to initiate and secure funding for this study. NRSB staff

---

[1]The committee's interim report is reproduced in Appendix A.

director Kevin Crowley was primarily responsible for starting the study. John Wiley, who served as study director, ably assisted the committee through all stages of information gathering, report development, and review. Staff members Toni Greenleaf, Darla Thompson, Marili Ulloa, Laura Llanos, and James Yates all helped bring this study to its successful conclusion.

David H. Leroy, Chair
Michael T. Ryan, Vice Chair[2]

---

[2]During the preparation of this final report Michael Ryan served as Chairman of the Nuclear Regulatory Commission's Advisory Committee on Nuclear Waste, which developed a white paper "History and Framework of Commercial Low-Level Radioactive Waste Management in the U.S." submitted to the Commission on December 30, 2005.

Contents

Overview

B y far the largest volumes of radioactive wastes in the United
States—millions of cubic meters—contain only small concentrations of radioactive material. These low-activity radioactive
wastes (LAW) should be regulated and managed according to their intrinsic hazardous properties and, thus, the degree of risk they pose for
treatment, storage, and disposal. The current regulatory structure is based
primarily on the wastes' origins[1] rather than their actual radiological risks.
There is no scientific basis for applying *different* degrees of control to
wastes that pose similar risks or applying *similar* controls to wastes that
pose very different risks. Such inconsistencies are inherent in the current
system.

In this report, the authoring committee[2] develops its vision of a risk-
informed system for regulating and managing all types of low-activity
waste in the United States. The framework for risk-informed decision
making combines scientific risk assessment with public values and perceptions. The framework is implemented in a gradual or stepwise fashion—
but always with regard to the hazardous properties[3] of the waste in

---

[1]The current system regulates LAW according to the enterprise that produced it (e.g.,
national defense, nuclear industry, nonnuclear industry, medicine).

[2]The National Academies Committee on Improving the Regulation and Management of
Low-Activity Radioactive Wastes. This study benefited from the support of eight domestic
and two international sponsors.

[3]While this report discusses explicitly only radiological hazards associated with low-
activity waste, the committee is well aware that these wastes often manifest chemical,

question and their comparison to those of other waste materials, and not to the enterprise that produced the waste.

The committee recognizes that public perceptions of risk may differ from scientific assessments. Determining a level of acceptable risk is a matter of public policy informed by science. The committee also recognizes the substantial body of laws and regulations and the large financial investment in management infrastructure, including disposal facilities, that are now in place. While regulatory authorities are adequate to ensure safety, the current system is complex, is inconsistent, and does not address risks of the various LAW systematically. The system is likely to grow less efficient in the future as more and different wastes are generated (e.g., from nuclear facility decommissioning, site cleanups, and new nuclear applications).

The committee found no easy way to reform the existing system. Although there has been some progress, efforts over the past 25 years to change the system generally have not been successful. Radioactive waste issues are highly controversial among citizens, especially those whose communities might be involved in waste facility siting or transportation routes. For public policy makers, the political liabilities for engaging in these issues are high and benefits are small. Nevertheless, among decision makers at all levels who are responsible for continuing to ensure the safety of LAW management, there is strong interest in improving current practices.

In addressing its charge, the committee sought to be practical. The report discusses and recommends a four-tiered system of change based on established principles for risk-informed decision making, current risk-informed initiatives by waste regulators in the United States and abroad, solutions available under current regulatory authorities, and opportunities for focused legislation as needed if simpler approaches are inadequate.

---

biological, and possibly other hazards. The risk-informed methodology developed in this report could, generally speaking, be extended to incorporate all such hazards, although the details of doing so are beyond the scope of this study. See Recommendation 1.

Summary

B y far the greatest volumes of radioactive wastes that arise annually in the United States contain only small concentrations of radioactive material. These low-activity wastes (LAW) present much less of a radiation hazard than either spent nuclear fuel or high-level radioactive waste. Improperly controlled, however, they have the potential to produce significant chronic (and in some cases acute) health risks. LAW arise in many sectors, including national defense, private industries, medicine, and research. Not all of these wastes are produced by enterprises that use nuclear materials or ionizing radiation—million cubic meter per year volumes arise incidentally in nonnuclear enterprises, primarily mineral mining and oil and gas recovery. These latter wastes contain naturally occurring radioactive materials (NORM), such as uranium, thorium, and their radioactive decay products, including radium and radon.

In the United States, LAW are subject to a regulatory patchwork that has evolved over almost 60 years. Statutes and regulations that control LAW are based primarily on the type of enterprise that produced it—the origin of the waste—rather than the waste's actual radiological hazard or potential health risk. The Atomic Energy Act of 1954 (AEA), as amended, provides federal control of nuclear energy-related enterprises, including their wastes. Federal control is exercised primarily by the Department of Energy (DOE), Nuclear Regulatory Commission (USNRC), and Environmental Protection Agency (EPA).

The Low-Level Radioactive Waste Policy Act of 1980 (LLRWPA), as amended, gave each state (or compacts of states) responsibility for disposing of a subset of AEA wastes, defined by statute as "low-level wastes," from private enterprises within the state. Generally speaking, the states control non-AEA wastes, such as NORM and TENORM[1] wastes. Both the USNRC and EPA have programs for withdrawing their federal authorities in order to allow the states to exercise their own authorities over public health and safety.

Private-sector enterprises and citizens are also important stakeholders in the management and regulation of LAW. Previous National Academies' studies found that disposing of slightly radioactive metal and concrete from decommissioning the current fleet of nuclear power reactors could cost $4.5 billion to $11.7 billion (NRC, 2002, p. 6) and that the cost of managing LAW is a major factor in biomedical research (NRC, 2001a). Citizens' perceptions of radiation risks can vary widely from those of technical experts, yet public perceptions of LAW are often important factors in decisions about disposal facility siting and waste transportation routes.

With this report, the committee[2] completes a two-part study to assess and recommend technical and policy options for improving practices for regulating and managing LAW (the statement of task appears in Sidebar 1.1). The committee finished the first part of its study with an interim report published in late 2003. The interim report, reprinted in Appendix A of this report, gives an overview of the current LAW system in the United States: waste characteristics, inventories, management and disposal practices, and federal and state regulations that control these wastes. In the interim report the committee found that there is adequate authority for managing LAW. However, the system is complex, and significant inconsistencies have arisen from regulating LAW mainly according to its origins rather than systematically considering its risks (see Sidebars 1.2 and 1.3).

In seeking ways to improve the system, the committee confronted the fact that current practices result from years of evolution of the origin-based system, involving many interactions among federal and state regulators, waste generators, and concerned citizens. Substantial change will not be easy. The objectives envisioned by Congress in the LLRWPA generally have not been met. Waste generators have only a limited number of disposal options, which often result in large volumes of waste being shipped long distances for disposal. The planned closure of the Barnwell,

---

[1]NORM that become more concentrated during mineral recovery or other operations are referred to as "technologically enhanced naturally occurring radioactive materials" (TENORM). TENORM includes material that has been made more accessible to human contact and therefore more likely to cause exposures.

[2]The Committee on Improving Practices for Regulating and Managing Low-Activity Radioactive Waste is referred to as "the committee" throughout this report.

South Carolina, site in 2008 could leave generators in more than 30 states without access to disposal for USNRC Class B and C low-level wastes. Significantly, however, federal regulatory agencies and other organizations have developed initiatives that could help improve the system. The DOE has developed an efficient strategy for disposing of very large volumes of very low activity wastes from its facility decommissioning and site cleanups. Chapter 2 summarizes these current initiatives and the near-term disposal situation.

To prepare this final report, the committee considered a number of options for improving the current system of LAW management. The committee came to the conclusion that a "risk-informed" approach would provide the best option for improving LAW regulation and management practices in the United States. A risk-informed approach is based on information provided by science-based risk assessment but includes stakeholders as a central component in decision making. Basing regulatory decisions and actions on the actual radiological hazards presented by the wastes themselves, and hence the risks they pose for their management and disposal, could provide the basis of a risk-informed framework for managing and disposing of the various types of LAW, and decisions within that framework would involve all stakeholders. The committee discusses these ideas in Chapter 3.

Another challenge for the committee was to agree how to move from the present origin-based system to a risk-informed system. Throughout its information-gathering activities, the committee heard a nearly unanimous opinion from congressional staff, regulators, generators, and public stakeholders that a sweeping conversion of the present origin-based patchwork of regulations and practices to a coherent system that uses risk as a basis for managing these wastes (i.e., a risk-informed system) would be most desirable (see Sidebar 4.3 of Appendix A). The same presenters, however, cautioned that such a conversion would be virtually impossible given the long history and investment in the regulatory and operational infrastructure of the current system, the disruption that an abrupt change could cause, and the lack of political will to effect such a change. Views varied widely about the urgency of changes and how to make them.

The committee found that while individual agencies and organizations are proposing important initiatives for moving toward an improved, risk-informed system, these single-agency initiatives lack priority. Better integration of these initiatives through cooperation among agencies could improve their chance of success. Integrated, practical, and stepwise improvements are most likely to succeed.

Chapter 4 describes a practical, tiered approach for making risk-informed changes under existing regulatory authorities, relying on congressional remedies when necessary. The committee distinguishes between the current "patchwork" approach of regulating, when the need arises, new or altered waste streams according to the enterprise that produced them, versus the committee's suggested "tiered" approach in which regulatory changes are directed toward controlling wastes according to their intrinsic radiological properties—with the appropriate level of control being determined through a risk-informed process in each instance.

Recommendation 1

The committee recommends that low-activity waste regulators implement risk-informed regulation of LAW through integrated strategies[3] developed by the regulatory agencies. Improving the system will require continued integration and coordination among regulatory agencies including the USNRC, EPA, DOE, DOD, and other federal and state agencies.

While current statutes and regulations for LAW provide adequate authority for protection of workers and the public, current practices are complex, inconsistent, and not based on a systematic consideration of risks. More efficient and uniformly protective management of the risks posed by these wastes will require moving away from the present origin-based regulatory system—a system that is firmly established through decades of practice and involves a number of federal and state agencies that have different authorities.

The development and use of integrated strategies would strengthen waste regulators' ongoing efforts to improve LAW regulation and management practices by

1. Focusing the attention of decision makers at all levels on the needs for and benefits of implementing risk-informed practices,
2. Providing a unified approach to developing risk-informed practices that is recognized by all stakeholders as cooperative and mutually supportive, and
3. Promoting harmonization (consistency on the basis of risk) in changes at each of the four tiers discussed in this report.

An important purpose of interagency strategies would be to help regulatory agencies balance their use of the four-tiered approach (see Rec-

---

[3]By "integrated strategies" the committee means the results of agencies working together to develop a single or joint strategies for using the approach in Recommendation 2 to implement risk-informed practices. Because the regulatory agencies have different legal authorities they may develop separate, but coordinated, strategies.

ommendation 2), including instances where targeted legislation[4] might be needed if the first three tiers are not sufficient for developing solutions. Cooperative interagency efforts have made significant progress in improving regulations in areas that are relevant to LAW management and disposal. Examples include development of the Multi-Agency Radiation Survey and Site Investigation Manual (MARSSIM)[5] and guidance from the Interagency Steering Committee on Radiation Standards (ISCORS), the latter of which includes eight federal agencies and has the goal of improving consistency in federal radiation protection programs. Development of the integrated strategies should build on the successes of MARSSIM, ISCORS, and similar interagency efforts and make even greater use of such efforts. Developing and instituting implementation strategies may require several years, as did the work on MARSSIM.

Two areas identified in this study exemplify where risk-informed regulations would improve the current system and could provide a focus for development of the strategies:

- Wastes containing uranium or thorium and their radioactive progeny generated by AEA- and non-AEA-controlled industries pose similar hazards (according to the type and concentration of their radioactivity) but are controlled under very different regulatory regimes.
- There is no generalized provision for wastes that contain very low concentrations of radioactivity to exit the regulatory system, although there are examples of case-by-case exemption or clearance of some such wastes.

Recommendation 2

The committee recommends that regulatory agencies adopt a risk-informed LAW system in incremental steps, relying mainly on their existing authorities under current statutes, and using a four-tiered approach: (1) changes to specific facility licenses or permits and individual licensee decisions; (2) regulatory guidance to advise on specific practices; (3) regulation changes; or if necessary, (4) legislative changes.

The committee advocates a stepwise "simplest-is-best" approach to implementing risk-informed LAW regulation and management. Acting under their existing authorities, regulatory agencies and site operators

---

[4]The 2005 Energy Policy Act's expanded definition of byproduct materials is an example of such legislation. See Chapter 2.

[5]See Chapter 4.

can effect significant changes from the bottom up, beginning with changes to specific facility licenses, permits, or decisions. By changing licenses and permits, the burden of moving toward risk-informed practices is shared by generators, facility operators, and regulators. Good business practices can lead generators toward better waste prevention, minimization, and segregation if there is more flexibility in selecting options for dispositioning their wastes. Chapter 4 provides details of these measures for implementing risk-informed LAW practices.

Recommendation 3

The committee recommends that government agencies continue to explore ways to improve their efforts to gather knowledge and opinions from stakeholders, particularly the affected and interested publics, when making LAW risk management decisions. Public stakeholders play a central role in a risk-informed decision process.

When those affected by a decision are involved in the decision-making process, the outcome is generally more accepted and more easily implemented than it would be otherwise. Management and disposal of LAW and other potential environmental hazards have evolved beyond ex post facto announcements by facility operators and regulatory agencies into a deliberative process involving partnerships with the affected and interested publics.

Several countries have been generally more successful than the United States in gaining public stakeholder support for siting LAW disposal facilities. As discussed in Chapters 3 and 4, reasons that these stakeholders have been more supportive include greater transparency of decision making, public enfranchisement and participation in decision making, better involvement of elected local officials, and ultimately the ability of local communities to veto an initial site selection. Besides outreach, another way a few government organizations in Europe and the United States have helped public stakeholders become more central in risk decision-making processes is by helping them hire their own technical experts.

While agencies with responsibility for LAW in the United States have improved their efforts to involve the public in waste disposal decisions, many citizens continue to perceive those efforts as falling short of their intended goals. A continuing, concerted effort is needed to understand and address those shortcomings and, in particular, ensure that public stakeholders are a central part of a risk-informed decision process.

Recommendation 4

The committee recommends that federal and state agencies continue to harmonize their regulations for managing and disposing of AEA and non-AEA wastes so that those wastes will be controlled consistently according to their radiological hazards rather than their origins.

In the interim report, the committee developed five categories that it considered inclusive of the spectrum of LAW and that helped to point out gaps and inconsistencies in present regulation and management practices. The two major deficiencies listed in Recommendation 1 stood out. The committee is not alone in recognizing these deficiencies. As discussed in Chapter 2, current initiatives by Congress, regulatory authorities, and other organizations are important initial steps in rectifying them. These initiatives should continue under current regulatory authorities as described in Chapters 2 and 4 and Recommendation 2.

Recommendation 5

The committee recommends continued collaboration among U.S. and international institutions that are responsible for controlling LAW. Greater consideration of international consensus standards as bases for U.S. regulations and practices is encouraged.

International organizations, especially the European Commission (EC) and the International Atomic Energy Agency (IAEA), are making significant progress in developing consistent, risk-based standards for managing LAW. Their approaches include a number of important elements of a risk-informed system. The IAEA waste classification system focuses on radiological properties of the waste rather than its origins. For example, at the very low activity end, EC regulations and IAEA standards provide guidelines for wastes to be cleared or exempted from control as radioactive material. At the high end, nuclear fuel reprocessing wastes and wastes with similar properties are classified as "high-level wastes." In the U.S. system, only wastes from reprocessing meet the legal definition of high-level waste, leaving other wastes that might pose similar risks to be defined as "greater-than-Class C low-level wastes," as discussed in Chapter 2.

Public stakeholders are likely to be more receptive to waste management practices that are known to be accepted and implemented in other developed countries. If waste management technical experts and regulators develop broad agreement, publics might be more trusting of their ability to ensure safe management and disposal practices. Moving toward risk-informed practices in the United States could have the net effect of increasing stakeholder support in all countries.

CONCLUSION

The committee concluded that, while challenging, it is possible to move in incremental steps to a more risk-informed system for controlling management and disposition of radioactive materials. In contrast with the patchwork evolution of the past 60 years, stepwise implementation would move in a consistent direction: away from regulating LAW according to how or when it was generated and toward regulation based on the actual hazard and potential risk of the material. Risk-informed practices are good business practices. By working with regulators, public authorities, and local citizens to implement risk-informed practices, industry can increase the cost-effectiveness of its LAW disposals and increase its options for such disposals; and by moving away from the ad hoc nature of the current origin-based system, industry can increase the predictability of its disposal options. Through open and objective dialogue, risk as perceived by generators, regulators, concerned citizens, and elected officials can provide a common basis—a common currency—leading to better cooperation, agreement, and progress.

1

Introduction

Wastes that contain small concentrations of radioactive materials arise from national defense, private industries, medicine, and research. Some of these wastes are produced by enterprises that use nuclear materials or ionizing radiation, while other wastes arise incidentally in non-nuclear enterprises such as natural resource recovery and water treatment. These low-activity wastes (LAW) are controlled by a regulatory patchwork that has evolved over almost 60 years to include a number of federal and state agencies. Laws, statutes, and regulations that control LAW are based primarily on the type of enterprise that produced it—the origin of the waste—rather than the waste's intrinsic radiological hazard.

LAW present less of a radiation hazard than either spent nuclear fuel or high-level radioactive waste. However, LAW may produce potential radiation exposure at levels above background that if not properly controlled may represent a significant chronic (and in some cases, an acute) hazard. For some LAW, the patchwork system of controls may be overly restrictive, providing only limited and expensive options for their management and disposal. On the other hand, the patchwork may result in the relative neglect and less control of other LAW that pose an equal or higher risk.

With this report, the committee[1] completes a two-part study to assess and recommend technical and policy options for improving practices for

---

[1]The Committee on Improving Practices for Regulating and Managing Low-Activity Radioactive Waste is referred to as "the committee" throughout this report.

> **Sidebar 1.1**
> **Task Statement**
>
> The objective of this study is to evaluate options for improving practices for regulating and managing low-activity radioactive waste in the United States. The study will focus on the following three tasks:
>
> 1. Using available information from public domain sources, provide a summary of the sources, forms, quantities, hazards, and other identifying characteristics of low-activity waste in the United States;
> 2. Review and summarize current policies and practices for regulating, treating, and disposing of low-activity waste, including the quantitative (including risk) bases for existing regulatory systems, and identify waste streams that are not being regulated or managed in a safe or cost-effective manner; and
> 3. Provide an assessment of technical and policy options for improving practices for regulating and managing low-activity waste to enhance technical soundness, ensure continued protection of public and environmental health, and increase cost effectiveness. This assessment should include an examination of options for utilizing risk-informed practices for identifying, regulating, and managing low-activity waste irrespective of its classification.

regulating and managing LAW. The committee finished the first part of its study with an interim report published in late 2003. The interim report addressed the first two items of the committee's task statement (see Sidebar 1.1) by providing an overview of LAW characteristics, inventories, management and disposal practices, and the federal and state regulations that control these wastes. The interim report is reproduced in Appendix A and summarized in the next section. Readers who seek background information on the topics discussed in this report should refer to Appendix A.

SUMMARY OF THE INTERIM REPORT

Federal authority for controlling nuclear materials dates back to the McMahon Act of 1946, enacted during the early period of development of nuclear weapons. Its successor, the Atomic Energy Act of 1954 (AEA), as amended, is the basis for today's federal control of nuclear energy-related enterprises and their wastes. However, a substantial portion of low-activity radioactive wastes comes from enterprises not regulated under the AEA, and their control devolves to state authorities.

Dividing the control of radioactive materials according to whether an AEA or non-AEA enterprise produced them began the evolution of the patchwork of regulations that today control LAW in the United States. Some of the more salient features of the origin-based patchwork are the following:[2]

• A large volume of AEA wastes falls under the statutory definition of *low-level waste* (LLW). The definition, however, is by exclusion—LLW are those not otherwise defined (e.g., as high-level waste, transuranic waste, or certain byproducts from uranium mining and milling).
• The Nuclear Regulatory Commission (USNRC), which regulates commercial uses of nuclear energy, subdivides LLW into Classes A (disposable with the least controls), B, and C that are deemed suitable for near-surface land disposal; and "greater-than-Class C," which currently has no disposal pathway.[3]
• Under the AEA, the Department of Energy (DOE) self-regulates LLW disposal at its own sites. DOE's orders for radioactive waste management and disposal are generally consistent with USNRC regulations, although DOE does not use the USNRC's subdivisions of LLW.
• Another large volume of AEA wastes results from the processing of uranium and thorium ores for nuclear energy applications. Generally speaking, wastes from the milling and extraction of uranium that were generated after 1978 are regulated by the USNRC, while pre-1978 wastes may be regulated by DOE, USNRC, or individual states.
• Regulations for non-AEA wastes, such as wastes that contain naturally occurring radioactive materials (NORM), vary considerably among the states. Nationwide, million cubic meter volumes of NORM wastes result from mining and oil and gas production each year.

To develop its interim report, the committee found it useful to take a step back from the present system of origin-based regulations and to look more closely at the wastes' radiological properties. This approach led the committee to divide the spectrum of LAW into five categories to serve as reference points for identifying and assessing options for improving the current practices (see Sidebar 1.2). The categories were not intended as a proposal for a new waste classification scheme but rather served the committee as a way to highlight inconsistencies between the wastes' radiological hazards and their regulation.

---

[2]See Appendix A for a detailed discussion of these and other relevant regulations as well as inventory data.

[3]Radioactivity concentration limits used by the USNRC are given in Table B.1 of Appendix A.

Sidebar 1.2
Interim Report's Overview of Inconsistencies in
Low-Activity Waste Regulations Versus Radiological Hazards

Wastes that fall within the legal definition of *low-level waste (LLW)* can have very different radiological properties:

1. Much LLW fits within the regulatory Classes A, B, and C.
2. However, large volumes of wastes from decommissioning and site cleanup often contain practically no radioactive material, but they cannot exit the regulatory system because Class A has no lower boundary.
3. Although defined as LLW, out-of-service radioactive sources can pose an acute exposure hazard, particularly if mishandled.

Other wastes that fall under different legal definitions can have very similar radiological properties:

4. Uranium and thorium mining and milling wastes are under federal control according to the AEA.
5. Wastes from the recovery of other natural resources or processes such as municipal water treatment can also contain uranium, thorium, and their progeny, but they are controlled by the individual states.

Although they are all legally defined as LLW, the wastes that comprised the committee's first three categories have very different radiological and physical characteristics. First of all, there are the wastes that fit appropriately into the USNRC classification system (e.g., Classes A, B, and C), such as those disposed at Barnwell (Chem-Nuclear/Duratek Disposal Systems), Hanford Washington (US Ecology), certain Class A wastes disposed at Clive Utah (Envirocare of Utah),[4] and wastes in typical DOE "burial grounds."

Second, there are the very large volumes of debris, rubble, and contaminated soils from DOE and commercial nuclear facility decommissioning and site cleanup that produce very low levels of radiation. They fall at the very low end of Class A but cannot exit the nuclear regulatory system[5] because the statutory definition of LLW has no lower boundary.

---

[4]On February 3, 2006, while this report was in press, a new company, EnergySolutions, was formed by Envirocare and two other companies. On February 7, 2006, EnergySolutions signed an agreement to acquire Duratek.

[5]Except by case-by-case exemptions discussed in Chapter 2.

> **Sidebar 1.3**
> **Committee's Findings in Its Interim Report**
>
> 1. Current statutes and regulations for low-activity radioactive wastes provide adequate authority for protection of workers and the public.
> 2. The current system of managing and regulating LAW is complex. It was developed under a patchwork system that has evolved based on the origins of the waste.
> 3. Certain categories of LAW have not received consistent regulatory oversight and management.
> 4. Current regulations for LAW are not based on a systematic consideration of risks.

Third, there are out-of-service radiation sources (often called "sealed sources") that typically contain pure or highly concentrated radioactive materials from industrial, medical, and research applications. They can emit levels of radiation sufficient to cause high individual exposures or lead to serious contamination incidents if they are improperly handled. Some of these sources exceed USNRC Class C but nevertheless meet the AEA definition of low-level waste.

The last two of the committee's five categories described in the interim report recognize the large volumes of wastes that contain uranium, thorium, and/or their radioactive decay products (progeny). Among these wastes, those that arise from the recovery of uranium and thorium for nuclear energy applications are legally defined as "byproduct" wastes in section 11e.(2) of the AEA. They are subject to federal control. Wastes that arise in mining, oil and gas production, coal burning, and other enterprises not related to nuclear applications can also contain uranium, thorium, and/or their progeny. These wastes are not included in the AEA, and at present their control is left principally to individual states.

Viewing the current LAW system in the context of these five categories led the committee to its findings that current statutes and regulations for low-activity radioactive wastes provide adequate authority for protection of workers and the public, but that the current system is complex, inconsistent, and does not address risks of the various LAW systematically (see Sidebar 1.3).

DEVELOPMENT OF THIS REPORT

The task of this final phase of the study was to assess technical and policy options to improve regulatory and management practices for LAW.

In seeking ways to improve the system, the committee confronted the fact that the present origin-based system is the product of years of evolution involving many interactions among federal and state regulators, waste generators, and concerned citizens.

The DOE has managed and disposed of large volumes of LAW at its own and commercial sites since the Manhattan Project. The Low-Level Radioactive Waste Policy Act, enacted in 1980 and amended in 1985 (LLRWPA), was a major effort by Congress to significantly revise private-sector practices by making each state (or regional compacts of states) responsible for disposing of its own commercially generated LLW.

The LLRWPA and other attempts to improve the system met with only limited success. Presently commercial waste generators have only a few disposal options that often result in large volumes of waste being shipped long distances for disposal. The planned closure of the Barnwell site in 2008, which could leave generators in more than 30 states without access to disposal for USNRC Class B and C wastes, is discussed in Chapter 2.

During the committee's open sessions, statements from public interest organizations and some members of the attending public expressed considerable lack of trust in the LAW regulatory system due to its complexity, inflexibility, and inconsistency. This lack of trust has apparently raised doubts among some members of the public about the current system's capability for protecting their health.

The committee, however, noted that there are few incentives for policy makers to become involved in LAW issues. Policy makers have necessarily focused their attention on high-level waste issues and the potential for misuse of nuclear materials (Wiley, 2005). The USNRC recently suspended rulemaking on alternative ways to disposition slightly radioactive materials citing these higher priorities. The 2005 Energy Policy Act extended AEA control of concentrated (discrete) NORM and accelerator-produced radioactive materials, but did not address low-activity (diffuse) forms of these materials. Nonetheless, there remain a number of important U.S. and international initiatives by regulators and other organizations that could improve current LAW practices. These initiatives are discussed in Chapter 2.

During the course of this study the committee came to the conclusion that a "risk-informed" approach would provide the best option for improving LAW regulation and management practices in the United States. A risk-informed approach is based on information provided by science-based risk assessment but includes stakeholders as a central component in decision making. Basing regulatory decisions and actions on the actual radiological hazards presented by the wastes themselves—rather than their origins—could provide the basis of a consistent framework for managing and disposing the various types of LAW, and decisions within that framework would involve all stakeholders. Chapter 3 begins with an overview of risk and the use of risk assessment, develops the concept of risk-informed decision making through broad stakeholder participation, and ends with the committee's vision of a risk-informed framework for regulating and managing LAW.

Another challenge for the committee was to agree how to move from the present origin-based system to a risk-informed system. Throughout its information-gathering activities, the committee heard a nearly unanimous opinion from congressional staff, regulators, generators, and public stakeholders that a sweeping conversion of the present origin-based patchwork of regulations and practices to a coherent system that uses risk as a basis for managing these wastes (i.e., a risk-informed system) would in principle be most desirable (see Sidebar 4.3 of Appendix A). The same presenters, however, cautioned that such a change would be virtually impossible given the long history and investment in the regulatory and operational infrastructure of the current system, the disruption that a sweeping change could cause, and the lack of political will to effect such a change.[6] Views varied widely about the urgency of changes and how to make them.

In its own discussions the committee focused on broad approaches to implementing risk-informed practices, well aware of the very contentious nature of radioactive waste issues in general, the apparent lack of progress in resolving these issues during the past 25 years or more, constraints imposed by the current laws and regulations for LAW, and federal and private investments in the present infrastructure. Options evaluated by the committee included sweeping legislative changes by Congress such as revision of the LLRWPA and the basic definition of LLW, adoption of internationally agreed-upon dose- or risk-based standards such as those developed by the International Atomic Energy Agency, and implementing changes under currently existing regulatory authorities.

Considering that the LLRWPA generally failed to meet the objectives envisioned by Congress and the perception of lack of political will to revisit such broadly targeted legislation (Leroy, 2004), the committee generally discounted the sweeping change option. Clearly there are opportunities where *specific* changes will require legislative action. The committee found important benefits from greater use of international standards and practices.

The committee concluded that components of each of the above options can and should be used to implement risk-informed LAW prac-

---

[6]Presentations at the committee's information-gathering meetings are listed in Appendix C.

tices. Chapter 4 describes a practical, tiered approach for regulatory agencies to make risk-informed changes under their existing authorities, relying on congressional remedies only when necessary. The committee distinguishes between the current "patchwork" approach of regulating, when the need arises, new or altered waste streams according to the enterprise that produced them, and the committee's suggested "tiered" approach in which regulatory changes are directed toward controlling wastes according to their intrinsic radiological properties—their appropriate level of control being determined through a risk-informed process in each instance.

Because implementing a risk-informed system is not the sole responsibility of regulators, Chapter 4 also describes responsibilities and opportunities for industry and public stakeholders in implementing the system. The concepts and approaches set out in Chapters 3 and 4 provide the basis for a developing an integrated LAW strategy, which the committee recommends in Chapter 5 along with other recommendations for improving the current system.

2

Current Initiatives for Improving Low-Activity Waste Regulation and Management

The committee's interim report presented an overview of current practices for regulating and managing low-activity radioactive wastes (LAW) in the United States.[1] This chapter extends and updates information presented in the interim report. The first section of this chapter describes initiatives by U.S. regulatory agencies and other organizations that are directed at improving the current LAW system. The second section summarizes international practices and initiatives for managing LAW.[2] The last section of this chapter addresses near- and longer-term issues regarding disposal capacity in the United States. Along with the interim report, the three sections of this chapter provide the basic picture of LAW regulation and management that led the committee to its views on how the present system might be improved.

CURRENT U.S. INITIATIVES

The interim report identified certain types of LAW that are not being managed efficiently under the present origin-based regulatory system.[3] Regulatory agencies, professional and commercial organizations, and members of Congress also have recognized deficiencies in the present system and have put forth several important initiatives to address them.

---

[1]The interim report is reproduced in Appendix A of this report.
[2]Appendix B gives a more detailed summary of international practices.
[3]See Sidebar 1.2 and also Chapter 4 of Appendix A.

This section describes these initiatives within the context of the types of LAW that the interim report described as posing management challenges:

- Slightly radioactive wastes that fall under the statutory definition of low-level waste (LLW),
- Highly concentrated radioactive wastes that are defined by statute as LLW,[4] and
- Wastes containing uranium- or thorium-series radionuclides, which are regulated inconsistently by federal and state agencies.

Slightly Radioactive Low-Level Wastes

A previous National Academies' committee reviewed disposition options for slightly radioactive solid wastes from decommissioning the nation's existing power reactors. That committee estimated costs of $4.5 billion to $11.7 billion for disposing of 10 million tons of concrete and metal debris in Nuclear Regulatory Commission (USNRC)-licensed LLW facilities (NRC, 2002, p. 6). For smaller enterprises with limited funds for waste disposal, finding a safe and economical disposal alternative can mean the difference between cleaning up a site and releasing it for unrestricted use, and leaving the waste in place or storing it until an affordable option becomes available (Federline, 2004).

This committee, along with the Environmental Protection Agency (EPA) and the USNRC as shown by their initiatives described below, considered whether other disposal methods may be able to provide protection for slightly radioactive wastes, given their low potential for posing radiological risks.

Low-Activity Waste Disposal in Landfills

In late 2003, EPA published an Advance Notice of Proposed Rulemaking (ANPR) describing the potential use of RCRA hazardous waste landfills[5] for the disposal of certain LAW, such as large-volume wastes that fall in USNRC Class A but are relatively low in radionuclide content (EPA, 2003). Subtitle C regulations require, among other things, that a

---

[4]Clearly these are not LAW. As discussed in this section, the committee included them to illustrate the shortcomings of statutory definition of wastes according to their origin rather than their actual radiological hazard.

[5]Hazardous wastes and their disposal are regulated by the EPA under Subtitle C of the Resource Conservation and Recovery Act (RCRA) of 1976, as amended. These landfills are described in the ANPR, which is available at *http://www.epa.gov/fedrgstr/EPA-WASTE/2003/November/Day-18/f28651.htm*.

disposal facility have a cap to minimize infiltration of liquids and a liner and leachate collection system beneath the waste. EPA received some 1500 public comments on the ANPR and is proceeding slowly in its rulemaking.

According to the ANPR, both EPA and USNRC believe that for certain wastes appropriate RCRA-permitted low-activity waste disposal can be as safe as disposal in USNRC-licensed facilities. EPA noted that RCRA landfills are currently being used for the disposal of a variety of radioactive wastes in accordance with state permitting requirements for these facilities. EPA's approach would establish a national framework for the regulation of these types of materials that would lead to more uniform regulation.

There are approximately 20 RCRA-permitted commercial disposal facilities in the United States, far more than the three commercial LLW disposal sites. Facilities in some states (e.g., Texas, Idaho) currently accept LAW exempted by the USNRC.[6] These facilities and others also accept some types of uranium-bearing wastes, which are discussed later in this chapter.

There are a few instances where states have permitted the use of RCRA Subtitle D municipal waste landfills for disposal of radioactive waste that contains very small concentrations of radioactive material. The committee noted in its interim report that very low activity wastes from the decommissioning of the Big Rock Point nuclear power plant were sent to a municipal landfill in Michigan. Other states, such as Texas, have determined that municipal landfills offer sufficient protection for certain types of radioactive material, for example, materials with very short half-lives, and have defined in their state regulations the kinds and amounts of radioactive wastes that may be so disposed.[7]

Limited or Free Release for Reuse

Since 1999 the USNRC has sought to develop a rule that would provide alternatives to disposing of slightly radioactive solid materials in licensed LLW facilities. On June 1, 2005, the commissioners of the USNRC disapproved the proposed rule "Radiological Criteria for Controlling the Disposition of Solid Materials" (USNRC, 2005), which had been prepared

---

[6]Under 10 CFR 20.2002 the USNRC has the authority to allow the release of very low level radioactive material from licensees, allowing disposal in unlicensed facilities on a case by case basis. The nuclear industry has found the 10 CFR 20.2002 process to be slow and expensive and, as a result, has submitted only about one alternate disposal application per year during the past 10 years (Genoa, 2003).

[7]Texas Administrative Code, Title 25, Chapter 289, Section 202(fff).

by the USNRC staff. The rule would have allowed disposal of some very low activity wastes at EPA-regulated RCRA landfills or conditional reuse of some materials (e.g., for roadbeds, bridges). The commission deferred further work on the rule due to higher-priority tasks as well as the previous National Academies' (NRC, 2002) finding that the USNRC's practice of case-by-case approvals of alternate dispositions is protective of public health.

In their individual comments, all of the commissioners indicated that such a rule needed further consideration, especially due to public stakeholder opposition to the proposed rule (see Chapter 4). Commissioner Merrifield commented, "This [rulemaking] is not just a simple matter of science. Recognizing the importance that our stakeholders place on this issue. . . . I felt that we needed to be a bit more creative in our approach to a complicated public policy issue" (Merrifield, 2005).

Bulk Waste Disposal

Because of its substantial efforts to clean up the Department of Energy's (DOE's) former nuclear materials production sites, DOE's Office of Environmental Management (EM) generates the nation's largest volumes of Atomic Energy Act (AEA) low-level wastes.[8] As noted in the interim report, DOE is responsible for managing and disposing of its own AEA wastes and regulates wastes at its sites according to DOE guidelines and orders. DOE wastes become subject to USNRC regulations only if they are shipped to a commercial LLW facility. The basic performance requirements in DOE guidelines and orders are generally consistent with USNRC regulations, although DOE does not use the USNRC classifications of A, B, C, and greater-than-class C wastes.

The Comprehensive Environmental Response, Compensation, and Liability Act (CERCLA), administered by EPA, applies to the cleanup of significantly contaminated sites, including the major DOE cleanup projects. Under CERCLA, DOE is required to follow a specified decision-making process, including public involvement, in planning site cleanup and waste disposal. A site-specific final plan is documented in a DOE record of decision, which EPA must approve.

DOE's policy is to dispose of its LLW at the generating site, if practical, or at another DOE site. In 2000, DOE designated Hanford, Washington, and the Nevada Test Site as sites that could receive LLW from all sites

---

[8]The committee did not examine EM's overall cleanup program or its disposal plans for other waste streams, such as transuranic or high-level wastes.

TABLE 2.1 Large-Volume Disposals of DOE LLW Through Mid-2005

Facility	Total Volume Disposed (m^3)	Comments
Fernald (on-site disposal)	2 million	Site will be closed in 2006
Hanford Environmental Restoration Disposal Facility (ERDF)	2.8 million	Remaining capacity estimated to be about 6.0 million cubic meters based on 10 cells. More cells could be built
Nevada Test Site	1 million	Current capacity is 3.6 million cubic meters; total capacity nearly unlimited[a]
Envirocare of Utah	1.2 million	About half of this DOE waste received from Fernald since 1999

[a]Becker et al. (2005).

SOURCE: DOE Office of Commercial Disposition Options.

throughout the DOE complex.[9] However, if DOE sites' own capabilities are not practical or cost-effective, DOE may approve the use of commercial treatment or disposal facilities.[10]

While DOE generates very large volumes of LLW, most consists of slightly contaminated soils and rubble from facility decommissioning and site cleanup. Table 2.1 provides an overview of how DOE has used a combination of its own and commercial disposal for these large volumes of slightly contaminated materials.

Fernald waste provides a good example of DOE's preference for using its own sites, but to use commercial disposal when necessary due to lack of on-site capability or when commercial disposal provides economic advantages. Because the Fernald site is being decommissioned and closed, DOE disposed on-site only the materials with the lowest concentration of radioactive material—mostly soils and foundations. The materials with a higher concentration of radioactivity, which still fell within USNRC Class A, were disposed in the commercial Envirocare facility. A last portion of the Fernald wastes, just under 7000 m^3, is being stabilized with cement (grout) for shipment and storage at the Waste Control Specialists (WCS)

---

[9]Importing out-of-state radioactive waste to Hanford has been challenged by the State of Washington. DOE had suspended imports of waste from other sites at the time this report was undergoing review.

[10]This activity is managed by DOE-EM's Office of Commercial Disposition Options, which provided the data on DOE waste disposal that are presented in this section.

commercial site in Texas. This waste, classified as 11e.(2) byproduct material, contains substantial amounts of ^{226}Ra. WCS is currently seeking a license from the State of Texas to permanently dispose of this waste.

In addition to disposing of very low level wastes in bulk at certain disposal facilities, the larger DOE sites each have disposal capability for wastes with concentrations of radioactive material comparable to USNRC Classes B and C. For example, Hanford disposed of about 2500 m^3 of containerized waste in its LLW facility in 2004, including about 760 m^3 from other DOE sites.

In summary, DOE is essentially self-sufficient for its own LLW disposal needs. The committee did not attempt to judge the extent to which DOE's LLW practices are risk-informed. However, in the broad perspective, DOE's LLW disposal initiatives are consistent with the committee's view of reasonable risk management practices: large volumes of wastes that present very little radiological hazard are disposed in relatively inexpensive bulk facilities. DOE's large disposal cells such as the Hanford's Environmental Restoration Disposal Facility are constructed under CERCLA and resemble RCRA Class C landfills—with cap, liner, and water collection system. Portions of DOE LLW that are radiologically more hazardous are disposed in DOE facilities comparable to USNRC-licensed LLW facilities. While there will continue to be challenges for DOE to implement cost-effective disposals of its many varieties of low-level wastes (GAO, 2005), DOE's LLW disposal practices contain important elements of a risk-informed system.

Highly Radioactive Low-Level Wastes

In examining wastes legally defined as LLW, the committee in its interim report noted that "low-level" does not describe the quantity or concentration of radioactive materials in LLW, rather it is an artifact of the Nuclear Waste Policy Act (NWPA).[11] The NWPA defines LLW to include all AEA wastes that are not subject to another statutory waste definition. There is no upper boundary on the concentration of radioactivity in LLW.

The regulations in 10 CFR Part 61 do, however, provide a system for classifying LLW based on the concentrations of radioactive materials present (Classes A, B, and C). Class B and Class C LLW can include reactor components, filters, ion exchange media, sludges, and radioactive sources

---

[11]Similarly, the National Council on Radiation Protection and Measurements noted, "The definition of low-level waste is particularly problematic. Contrary to the common meaning of 'low-level' . . . low-level waste can contain high concentrations of shorter-lived and longer-lived radionuclides similar to those in high-level waste. The definition . . . may foster mistrust by the public because the simple question of what low-level waste is cannot be given a direct answer" (NCRP, 2002, p. 16).

Sidebar 2.1
Use and Disposition of Radioactive Sources

In the early 1900s, radioactive sources (particularly using radium) were introduced in industrial, medical, and research applications. In the middle of the century as man-made radioisotopes became increasingly available, the distribution and use of radioactive sources became widespread. Today radioactive sources are in use worldwide. Unfortunately, while the applications of these sources have expanded rapidly, until recently detailed planning has not been given to their eventual disposition.

After a radioactive source has reached the end of its useful life, maintaining control of it or disposing of it are both expensive options since it is likely to still be highly radioactive. Often it is not clear when the source has truly become waste as opposed to simply being of no more use to the owner. Eventually, some of these sources end up being abandoned through one mechanism or another (e.g., controls are reduced and eventually terminated, records and chain of custody are lost or forgotten).

Examples of radioactive sources that can produce serious radiation exposures or contamination events if abandoned include brachytherapy sources (^{137}Cs, ^{192}Ir), sources for well logging and mobile industrial radiography (^{137}Cs, ^{192}Ir, ^{60}Co, ^{169}Yb, ^{170}Tm, ^{75}Se), radiothermal generators (^{90}Sr, ^{238}Pu), moisture gauges and static electricity "preventers" (^{226}Ra, ^{210}Po), and neutron generators (^{241}Am-Be).

---

used in medicine and industry (discrete sources). By volume these wastes comprise only a small fraction of the LLW inventory while containing more than 90 percent of the radioactive material (see Chapter 3 of Appendix A).

Control of Orphan Radioactive Sources

National and international initiatives are in place to control, recover, and properly dispose of orphaned[12] and no longer useful radioactive sources. Some of these sources can pose acute risks to the public and the environment (see Sidebar 2.1).

In the United States, the Department of Energy operates the Off-Site Source Recovery Program (OSRP) to recover and store certain excess and unwanted radioactive sources that potentially pose threats to national

---

[12]EPA uses the term "orphan" to refer to sources for which no owner can be identified.

security. This program operates under the U.S. Radiological Threat Reduction Program of the National Nuclear Security Administration.[13] This initiative is a key component of the DOE Global Threat Reduction Initiative created in May 2004. While the sources of major concern are those that represent a security threat within the United States, the program maintains, on an interim basis, its original concern only with sources that are greater-than-Class C (GTCC) in radionuclide content. The OSRP focuses on recovering ^{241}Am, ^{238}Pu, ^{239}Pu, ^{252}Cf, ^{244}Cm, ^{137}Cs, ^{90}Sr, ^{60}Co, ^{192}Ir, and ^{226}Ra sources. As of the end of fiscal year (FY) 2004, this program had recovered more than 10,500 sources from industrial sites, schools, universities, hospitals, and research institutions in almost every state.[14] The program is now focusing on higher-risk sources of about 200 Ci or greater. The recovered sources are intended to be recycled for other uses if possible. Otherwise, they will be stored or eventually disposed. The OSRP began supporting International Atomic Energy Agency (IAEA) source recovery efforts in FY 2005.

The Conference of Radiation Control Program Directors (CRCPD) assists states in retrieving and disposing of radioactive sources through its Orphan Sources Initiative. Through this initiative, the CRCPD and EPA enlist the participation of states, the USNRC, and the DOE in developing a nationwide program for controlling orphan sources. In certain limited cases, the EPA and DOE, through CRCPD, provide funds to state radiation control programs for the disposition of radioactive sources when the owner cannot afford the costs of disposition or should not be held liable for those costs. The CRCPD also offers assistance in finding affordable, legal disposition mechanisms, identifies contacts with appropriate government agencies, identifies other entities that may have a use for the source, and supports the OSRP.[15]

While discrete sources have attracted attention based upon potentially adverse consequences of illegitimate use, their actual hazards depend upon the concentration and total quantity of their radioactive material, its decay properties, its chemical and physical form, and its container. Many radioactive sources pose little, if any, threat to human health or the environment if properly disposed (see Sidebar 2.2).

Greater-than-Class C Low-Level Radioactive Waste

LLW that contains concentrations of radioactive material that exceed USNRC Class C can include discrete sources, reactor components, and

---

[13]See http://osrp.lanl.gov.
[14]See http://osrp.lanl.gov.
[15]See http://www.crcpd.org/orphans.asp.

Sidebar 2.2
Risk-Informed Discrete Source Disposal

In practice, practical determinations are characteristic of a risk-informed approach to regulating small, concentrated sources of radioactive material because the total amount of radioactive material, the concentration of the material, and the robustness of its disposal package are all considered as part of a risk-informed decision. Other aspects of a risk-informed decision-making process, such as political, economic, and stakeholder issues, also would enter into consideration.

For example, the requirements for disposal at the LLW facility in Barnwell, South Carolina, include limiting the curie strength of the sources for short-lived radionuclides and encapsulation of discrete sources in concrete. The maximum quantity for ^{137}Cs or ^{90}Sr is 25 Ci in each encapsulation. Very concentrated sealed sources (e.g., ^{90}Sr sources used in ophthalmology) may, on a curie-per-cubic-centimeter basis, be greater than Class C LLW. However, such sources can be disposed as LLW because their total curie content is small. The encapsulation concrete must be at least 4 inches thick with a minimum compressive strength of 2500 pounds per square inch. The containers are typically 30-gallon or 55-gallon steel drums. This scheme provides a more robust and predictable disposal package. Short-lived-radionuclide sealed sources can be mixed in the same encapsulation container up to a total of 25 Ci. Tritium gaseous sources must be packaged in a high-integrity container with each container limited to 1000 Ci. There is no specific curie limit on ^{60}Co for classification; however, the limiting factor is the radiation limit on the package to meet transportation requirements.

SOURCE: Personal communication from William B. House, Chem-Nuclear Systems, Barnwell, South Carolina to committee member Michael T. Ryan.

contaminated equipment. In a notice published in the *Federal Register* on May 11, 2005, DOE announced its intent to prepare an Environmental Impact Statement (EIS) for the disposal of GTCC LLW, pursuant to the Low-Level Radioactive Waste Policy Amendments Act of 1985. DOE's goal is to issue a Notice of Intent in mid-summer 2006, then complete the EIS within one and a half to two years. A progress report is due to Congress by August 8, 2006. The EIS is expected to evaluate the environmental impacts of disposal methods (e.g., enhanced near surface, greater confinement disposal, deep geologic repository) as well as locations for the disposal of the waste. DOE wastes with characteristics similar to GTCC

LLW that otherwise do not have a path to disposal may also be included in the scope of the EIS.

Wastes Containing Uranium- or Thorium-Series Radionuclides

In its interim report, the committee recognized that some of the large volumes of wastes containing uranium, thorium, and their progeny[16] date back to the Manhattan Project, when uranium was first mined and processed for the nuclear weapons program. More recently these wastes have resulted from both defense and civilian nuclear uses of uranium (see Appendix A). Uranium mining wastes are excluded from the AEA, but waste from milling uranium ore for nuclear energy applications is federally controlled by the AEA, as amended by the Uranium Mill Tailings Radiation Control Act of 1978 (UMTRCA). This means that the authorities of the USNRC, which are derived from the AEA, extend only to the mill tailings in a waste impoundment. States may have jurisdiction over the radioactive constituents in mining wastes, and the EPA may delegate its authority for the chemically hazardous constituents in mining wastes to states.

Responsibility for the uranium milling wastes controlled by the AEA passed through several federal agencies until UMTRCA was enacted in 1978. UMTRCA facilities are subject to EPA's standards in 40 CFR Part 192,[17] which are implemented by USNRC's regulations in 10 CFR Part 40. The USNRC's regulations are also based in part on EPA's RCRA hazardous waste standards. UMTRCA includes specific provisions for, among other things, radiation protection, radon mitigation, and long-term care and ownership by the DOE or the state in which the facility is located, with USNRC regulatory oversight of the long-term care.

Because uranium, thorium, and their radioactive progeny exist naturally on Earth, they are also found in wastes from enterprises, including mineral recovery and water treatment, that are not subject to the AEA (see Table 3.2 of Appendix A). States have general regulatory authority to protect the health and safety of their populations, and regulating "naturally occurring radioactive materials" (NORM) and TENORM [18] is one area in

---

[16]"Progeny" are the isotopes that result from the radioactive decay of "parent" isotopes. Progeny of uranium and thorium are themselves radioactive (see Table 3.1 of Appendix A).

[17]*Standards for Cleanup of Land and Buildings Contaminated with Residual Radioactive Materials from Inactive Uranium Processing Sites*, Subpart B of 40 CFR Part 192 (48 FR 590 to 606).

[18]NORM that become more concentrated during mineral recovery or other operations are referred to as "technologically enhanced naturally occurring radioactive materials" (TENORM). TENORM includes material that has been made more accessible to human contact and therefore more likely to cause exposures. Simply for convenience this report frequently uses the acronym NORM to include NORM and TENORM wastes (see also Sidebar 3.2 of Appendix A).

which some states have developed more detailed rules and regulations than others in asserting this authority. EPA has authority to regulate NORM under several statutes, including the Clean Air Act; the Toxic Substances Control Act; CERCLA; and RCRA (see Sidebar 2.3).

Generally speaking, NORM wastes have received little public attention although they are a significant contributor to background exposure in the United States (NRC, 1999). Management of NORM may be less controlled relative to some AEA LAW. Generators of significant amounts of these wastes include coal-burning power plants, oil and natural gas production, and water treatment plants. Eventually, risk-informed regulations may lead the states and federal agencies to require all such industries to characterize these materials by their radioactivity content and dispose them in an approved fashion, with radioactive materials content being one factor taken into account. The CRCPD has taken a significant step toward a regulatory framework for NORM wastes by developing suggested state regulations for these wastes and a guide to implementing the regulations.[19]

Wastes from the Formerly Utilized Sites Remedial Action Program (FUSRAP)

On legal grounds, the USNRC determined that it has no authority over uranium milling wastes at sites that were not licensed by the commission before UMTRCA was enacted.[20] Pre-UMTRCA wastes that are not subject to federal regulation under the AEA are subject to regulation under state authorities. Thus, essentially identical wastes are or are not subject to USNRC control depending only on when they were generated.

The formerly utilized sites remedial action program is managed by the Army Corps of Engineers (see Sidebar 3.1 of Appendix A). FUSRAP wastes amount to 1 to 2 million cubic meters of material—mainly soils containing uranium, thorium, and their progeny. Concentrations range from background (1-3 pCi/g) to approximately 10 times average background values. These wastes are excavated and shipped for disposal. The Corps estimates that some 80 percent of the wastes are pre-UMTRCA, with 20 percent being post-UMTRCA. Absent regulation by the USNRC, the states can, and some do, exercise control over the pre-UMTRCA waste. This has led to instances of inconsistent control that increase costs of FUSRAP cleanups, require transporting large amounts of very slightly contaminated wastes over long distances, and cause friction between

---

[19]See http://crcpd.org/SSRCRs/TOC_8-2001.htm.

[20]Except for wastes at Title I sites licensed by the USNRC after cleanup by DOE. Title I of UMTRCA provides that USNRC license these pre-UMTRCA sites after DOE has disposed of the tailings on them in accordance with EPA's standards at 40 CFR Part 192.

Sidebar 2.3
TENORM Wastes from Phosphate Mining:
An Example of EPA Control of Low-Activity Wastes

The principal resource for chemical fertilizer is calcium fluoro-phosphate (mineral name, collophane). Major production in the United States comes from Florida. Processing these deposits involves froth flotation to produce a concentrate containing at least 30 percent P_2O_5 followed by dissolution of the concentrate with sulfuric acid to form phosphoric acid.

In the flotation step, a solids size separation is made at 104 μm. The solids finer than 104 μm in size are added to tailings impoundments. The material that falls in the size range of 104 to 417 μm is subjected to froth flotation. The tailings from this operation are pumped to mine cuts for reclamation. There are an estimated 1.5 billion tons of tailings in the tailings impoundments in Florida, and 30 million tons are added each year. The solids in these impoundments contain about 40 pCi/g of radioactivity in the form of radium and uranium.

The waste product from phosphoric acid production is phosphogypsum, a solid material that contains calcium phosphate and sulfate. Approximately one billion tons of phosphogypsum are impounded in very large mounds, referred to as stacks, and 30 million tons are added annually to these stacks. This material contains about 30 pCi/g of radioactivity, primarily as radium sulfate.

Breaching of the gypsum stacks and tailing dams is a concern. In the summer of 2004, a phosphogypsum stack breach released millions of gallons of highly acidic water. The breach was contained rapidly, and significant environmental damage was not observed. In the case of the phosphogypsum stacks, there is a plastic liner underneath the stacks, and any runoff is collected.

Both radium and uranium go to the concentrate. During phosphoric acid production, greater than 90 percent of the uranium remains in the phosphoric acid solution, while radium (as radium sulfate) goes to the phosphogypsum waste. There is no practical way to separate the radium from the phosphogypsum. However, uranium was recovered from the phosphoric acid before the fall in the price of uranium in the 1980s. With the present price of uranium, discussion is under way to once again recover and remove the uranium from the phosphoric acid.

EPA radioactive waste disposal standards for phosphogypsum stacks have been adopted for use internationally. Greece's Atomic Energy Commission, Department of Environmental Radioactivity, recently developed standards for disposal of phosphogypsum into stacks that were closely based on EPA's requirements promulgated under the U.S. Clean Air Act for limitations on public exposures to radon (40 CFR 61, Subpart R). Disposal of phosphogypsum in stacks has become a common practice in other countries such as Canada, Spain, and Brazil.

federal and state agencies (McDaniel, 2004). Envirocare of Utah has disposed of a large portion of FUSRAP waste, while U.S. Ecology, Idaho, is currently providing most of the disposal capacity (see Chapter 4).

FUSRAP wastes are good examples of wastes that are currently being managed and disposed according to complex and often inconsistent regulations that have no clear relation to the wastes' radiological hazards. In comparing FUSRAP and NORM wastes, it is notable that the *annual* production of NORM wastes is about the same as the *total* volume of FUSRAP wastes, and the radiological hazards of NORM and FUSRAP wastes are comparable. Estimates of the total cost for disposal of FUSRAP are approximately $2 billion.[21]

NORM and Other LAW Disposal in UMTRCA Mill Tailings Impoundments

In 2004, the National Mining Association and the Fuel Cycle Facilities Forum submitted a white paper for consideration by the USNRC that proposes UMTRCA impoundments as a potential disposition path for NORM and other low-activity materials.[22] The white paper argues that the USNRC's 10 CFR Part 40, Appendix A criteria, including tailings impoundment design and site closure requirements, can ensure safe containment of a wide range of potential radiological and/or chemically hazardous nonradiological wastes in addition to those defined by statute as 11e.(2) wastes. Such wastes could include depleted uranium- and thorium-contaminated wastes, radium-contaminated TENORM wastes, and some special nuclear material-contaminated wastes. The paper notes that while RCRA disposal sites have a post-closure regulatory horizon of 30 years using active controls, 10 CFR Part 40 requires passive controls for tailings impoundments for a period of at least 200 years and, to the extent practicable, 1000 years.

According to the proposal, while some categories of candidate waste materials already may be accepted for direct disposal (e.g., source material), others should be equally acceptable if they pose similar risks. The white paper proposes the use of memoranda of understanding or other regulatory agreements between USNRC and EPA or other relevant regulatory agencies to mitigate or eliminate potential regulatory obstacles (e.g., dual or overlapping regulation) to the expanded use of uranium mill tailings impoundments. The white paper concludes that uranium mill tailings impoundments offer a direct disposal alternative that adequately protects public health and safety from the potential radiological

---

[21]See http://www.ead.anl.gov/project/dsp_fsdetail.cfm?id=61.
[22]See http://www.nma.org/pdf/legal/white_paper_non11e2submission_052804.pdf.

and nonradiological risks associated with many non-11e.(2) byproduct material waste streams.

Federal Control of Discrete TENORM Sources

In early 2005, the Health Physics Society and the Organization of Agreement States proposed congressional action to put concentrated ("discrete") TENORM sources—especially radium sources—and radioactive materials from particle accelerator operations under the AEA. These groups recognize that consistent, uniform regulation of all radioactive materials is needed, especially for sources that present significant radiation hazards and could potentially be used as "dirty bomb" material.[23] Uniform federal regulation of accelerator-produced radionuclides was also sought by the radiopharmaceutical industry.

The USNRC included a similar proposal in a suggested draft bill to amend the AEA. The transmittal letter to the Senate listed 11 objectives for the proposed legislation, one of which is "augmentation of the Commission's regulatory authority to protect the public health and safety and promote the common defense and security with respect to radioactive materials by including accelerator-produced and certain other radioactive material under its jurisdiction" (Diaz, 2005a, p. 1). The proposed bill sought to achieve this objective by revising the definition of *byproduct material* that is subject to the USNRC's AEA jurisdiction.

As this report was being finalized for review, the proposed legislation was incorporated into the Energy Policy Act of 2005, which was enacted on August 8, 2005.[24] The act imposes an aggressive schedule for the USNRC to issue final regulations by February 7, 2007. Although work on these regulations, including the definition of the term "discrete source," is only beginning, it seems clear that placing these materials under the AEA is an important step toward making their control more uniform and consistent with their actual radiological properties and risks.

INTERNATIONAL INITIATIVES

In moving toward more consistent, risk-informed management of LAW in the United States, the committee sees opportunities for greater exchange of ideas with the international community. Such exchanges could mutually enhance the knowledge of those responsible for LAW and their credibility within each country.

---

[23]See http://www.hps.org/documents/MaterialControl.pdf.
[24]See http://energycommerce.house.gov/108/energy_pdfs_2.htm.

FIGURE 2.1 The Morvilliers, France, site (foreground) is the world's first facility designed especially for disposing of very low activity radioactive wastes. Low- and intermediate-activity short-lived wastes are disposed of at the Centre de l'Aube (background). These facilities are located about 250 kilometers east of Paris.

Photo courtesy of P. Bourguignon, Agence nationale pour la gestion des déchlets radioactifs (ANDRA), France.

France recently opened a disposal facility for large volumes of very low activity wastes at Morvilliers (see Figure 2.1, foreground). This facility is physically separate from the nearby Centre de l'Aube facility (see Figure 2.1, background), which is designed for the relatively smaller volumes of wastes that are more typical of the USNRC Class A, B, and C wastes. The disposal trenches at Morvilliers are similar to EPA hazardous waste landfills, including a trench cap, liner, and leachate collection system. Spain has recently begun operating special disposal cells for very low activity wastes at its El Cabril facility. The cells are constructed according to hazardous waste requirements (Zuloaga, 2003). Japan has special regulations for very low level waste from its nuclear industry and is considering regulations for other types of LAW (Hirusawa, 2004). In parallel with these considerations, a risk-informed system is recommended by the Nuclear Safety Commission of Japan as an attempt to establish a unified framework for all types of radioactive wastes (NSC, 2004).

The international community has made significant progress toward establishing a consistent risk-based framework for managing radioactive wastes. As described below, the framework rests on dose-based standards developed by the IAEA and the International Commission on Radiological Protection (ICRP) to protect workers and the public from ionizing radiation. These standards are incorporated in the European Commission's directive 96/29 (EC, 1996a), which ensures consistency in protecting the public and workers from potential exposures to radiation, including those associated with waste management, in the European Union's 25 member countries (see also Appendix B).

Standards for Radiation Protection

The International Basic Safety Standards for Protection Against Ionizing Radiation and for the Safety of Radiation Sources (BSS) promulgated by the IAEA are a worldwide reference for protection from radiation (IAEA, 1996). Key concepts in the BSS include

1. *Exclusion* from regulation of exposures that are not amenable to control, for example exposure from ^{40}K in the body, cosmic radiation, and unmodified concentrations of radionuclides that are present in most raw materials;
2. *Exemption* of practices or materials resulting from those practices that do not require radiation protection and therefore never enter the regulatory system; and
3. *Clearance* of slightly radioactive materials, which allows them to be removed from regulatory control.

Practices that might produce radiation exposures must be justified, and further, they must be optimized to ensure that radiation exposures are kept as low as reasonably achievable (ALARA)—this is the same principle used by the USNRC and DOE in their regulations and orders. The BSS recommends that effective doses incurred from normal practices involving radioactive substances not exceed 20 millisieverts (mSv)/year (averaged over five years and not exceeding 50 mSv in a single year) for the worker, and 1 mSv for the relevant critical groups of the public.[25] The BSS also provides the basis for international control of radioactive sources, which present a growing security concern worldwide (see Sidebar 2.4).

---

[25]See Sidebar 3.1 for an explanation of these dose units. U.S. standards are substantially the same as the BSS, although numerical differences between some USNRC and EPA standards continue to be the subject of discussion between these agencies and among public stakeholders.

Sidebar 2.4
International Initiatives for Controlling Radioactive Sources

Even before the events of September 11, 2001, there was international recognition of the hazards presented by the lack of control of radioactive sources. Serious accidents, such as the ^{137}Cs contamination in Goiania, Brazil, in 1987 (IAEA, 1988), had demonstrated just how deadly such sources could become. In 1996, the IAEA and five other international organizations issued the BSS that established general requirements for the safety and security of radioactive sources (IAEA, 1996). In 1998 at the International Conference on the Safety of Radiation Sources and the Security of Radioactive Materials, a basis for a coordinated international approach to the safety of such sources was established. This was followed by conferences in 2000, 2001, 2003 (with post-9/11 emphasis), and June 2005 (IAEA, 1998, 2000, 2001, 2003a, 2005b). These conferences focused on the security of sources, the responsibilities of senior regulators in dealing with these matters, sustainable infrastructures for the control of radioactive sources, and methods for identifying, locating, and decommissioning orphaned sources. The IAEA *Code of Conduct on the Safety and Security of Radioactive Sources* (IAEA, 2004a) and *Guidance on the Import and Export of Radioactive Sources* (IAEA, 2005a) represent the culmination of these and similar efforts to provide guidance on how IAEA member countries can safely and securely manage radioactive sources that pose significant risk. The U.S. Energy Policy Act of 2005 uses the IAEA Code of Conduct's categorization of radiation sources (Conference Report on H.R. 6, section 170H).

Publication 81 of the ICRP provides guidance for controlling potential long-term exposures from waste disposal (ICRP, 1998). Recognizing that potential doses to future populations from waste disposal can only be estimated, ICRP recommends control by "constrained optimization" rather than by imposing specific dose limits. Optimization, according to ICRP 81, is a judgmental process with social and economic factors being taken into account, which should be carried out in a structured but essentially qualitative way. For waste disposal, constrained optimization includes meeting a dose constraint during normal operation of the disposal facility, reducing the likelihood or the consequences of inadvertent human intrusion, and using sound engineering and management practices in implementing disposal.

Regarding the dose constraint, ICRP recommends that estimates of doses to future populations not exceed 0.3 mSv under normal conditions (consistent with the BSS limit of 1 mSv, but considering possible exposures from other nuclear applications). In addition to dose constraint in the normal long-term performance of the disposal facility, ICRP recommends referring to two values of dose incurred in case of inadvertent intrusion into the disposal site: 10 and 100 mSv. According to ICRP, intervention to reduce radiation exposure is seldom required below 10 mSv and almost always required for doses above 100 mSv. Most countries use dose constraint rather than risk constraint in their national regulations. Even in Sweden and the United Kingdom, where the standard is expressed in terms of risk, the actual regulation is expressed in terms of dose.

ICRP recognizes that instead of using dose constraints, similar levels of protection can be achieved by using a risk-based approach (integrating dose estimates with probability, see Chapter 3). The 0.3 mSv per year constraint for waste disposal optimization in ICRP 81 is only a factor of 2-3 above the EPA risk criterion of 10^{-4} lifetime risk. ICRP also recognizes a two-pronged dose-probability approach in which the likelihood of exposure and the dose estimates are evaluated separately. ICRP considers that the latter approach allows obtaining more information for purposes of decision making.

European Commission directive 96/29 enforces the IAEA and ICRP recommendations on justification of practices, exemption, optimization, and dose constraint. The directive covers all activities involving radioactive material. It also addresses the possibility of enhanced exposure to natural radiation resulting from nonnuclear activities. Through its consistency with the IAEA and ICRP standards, the European Union has taken a major step toward establishing a unified system for radiation protection that covers waste management, including the disposal of waste originating from nuclear as well as nonnuclear industry. Detailed aspects remain to be worked out, some of which are noted in Appendix B.

IAEA Waste Classification

In 1994 the IAEA recommended an international waste classification system that generally reflects the radiological characteristics of wastes rather than their origins (IAEA, 1994). The basic classification system does not distinguish between radioactive wastes from the nuclear fuel cycle and from non-fuel-cycle wastes, such as NORM. For example, high-level waste (HLW) in the IAEA system includes *all* wastes with radioactivity levels similar to wastes from nuclear fuel reprocessing (see Table 2.2). As noted previously, the origin-based U.S. definition of HLW includes *only* fuel reprocessing wastes—leaving non-reprocessing wastes that are highly radioactive to be classified as "greater-than-Class C low-level waste." Concentrations of shorter-lived radionuclides in low- and intermediate-

TABLE 2.2 IAEA Waste Classification System

Waste Classes	Typical Characteristics	Disposal Options
1. Exempt Waste (EW)	Activity levels at or below clearance levels given in IAEA (2004b), which are based on an annual dose to members of the public of less than 0.01 mSv	No radiological restrictions
2. Low- and intermediate-level waste (LILW)	Activity levels above clearance levels and thermal power below about 2 kW/m^3	
2.1 Short-lived waste (LILW-SL)	Restricted long-lived radionuclide concentrations (the long-lived alpha-emitting radionuclide concentration in individual waste packages is limited to 4000 Bq/g, with the overall average for all packages limited to 400 Bq/g)	Near-surface[a] or geological disposal[b] facility
2.2 Long-lived waste (LILW-LL)	Long-lived radionuclide concentrations exceeding limitations for short-lived waste	Geological disposal facility
3. High-level waste (HLW)	Thermal power above about 2 kW/m^3 and long-lived radionuclide concentrations exceeding limitations for short-lived waste	Geological disposal facility

[a]IAEA (2003b).
[b]IAEA (2003c).

SOURCE: IAEA (1994).

level wastes are limited by the amount of decay heat they generate (thermal power density). There is no such restriction on U.S. LLW (i.e., the NWPA provides no upper limit on its definition of LLW).

Significantly, the IAEA system includes classes of materials that can be exempted from radiological controls or can be released (cleared) for disposal without radiological restrictions. A dose-based standard is used to determine materials that can be cleared or exempted.

International regulations are converging on a value of 1 mSv of added annual dose to the public as an appropriate limit for normal exposures arising from applications of radioactive materials, including waste management practices. The USNRC and DOE also use 1 mSv per year as the limit for doses to the public (DOE, 2005). There remain significant differences in applying ALARA among countries, ranging from a technological approach of designing as effective a confinement system as possible to a fully integrated risk-based probabilistic approach. Whatever approach to ALARA is taken, what constitutes "reasonable" may differ based upon the perceptions of the various stakeholders and concerned parties. Resolution of these differences requires genuine participation of divergent view holders in the decision-making process.

THE CURRENT U.S. DISPOSAL SITUATION AND POST-2008 ISSUES

The committee's interim report noted that no new LLW disposal sites have been developed by the states or interstate compacts as intended by Congress in the Low-Level Radioactive Waste Policy Act of 1980. The nation's only facilities licensed to dispose of LLW are located near Barnwell, South Carolina; Clive, Utah; and Richland, Washington. These sites were established and continue to be operated by private-sector companies,[26] subject to laws and requirements of their host state (all host states are USNRC Agreement States) and the state's regional compact. Two sites currently accept wastes from generators nationwide—the South Carolina site is licensed to accept all classes of LLW (USNRC Classes A, B, and C), and the Utah site is licensed to accept only Class A wastes. The Washington site accepts all classes of LLW, but only from states that are members of the Northwest and Rocky Mountain Compacts.[27] At the beginning of 2005, WCS, a private company seeking to develop a new disposal facility near Andrews, Texas, applied to that state for a Class A, B, and C license. The license application is under regulatory review and could be granted by 2007 if the review process proceeds without an interruption or delay in schedule (Jablonski, 2004).

In 2001, South Carolina enacted legislation to close the Barnwell site in 2008 to states outside the Atlantic Compact.[28] This action could leave generators in more than 30 states without access to disposal for their Class B and C wastes and dependent on the Utah site for disposal of their

---

[26]Chem-Nuclear Services/Duratek operates the Barnwell site; Envirocare of Utah operates the Clive site; and U.S. Ecology operates the Richland site. On February 3, 2006, while this report was in press, a new company, EnergySolutions, was formed by Envirocare and two other companies. On February 7, 2006, EnergySolutions signed an agreement to acquire Duratek.

[27]The state compacts and their members are listed in Table 2.1 of Appendix A.

[28]The Atlantic Compact consists of Connecticut, New Jersey, and South Carolina.

Class A wastes. Some view this as a serious consequence of the failure of the states and state compacts to develop even one new disposal facility (Leroy, 2004; Meserve, 2005; Pasternak, 2003). Others, including the Government Accountability Office (GAO) and the LLW Forum,[29] note that there is abundant capacity for Class A waste and the relatively small volumes of Class B and C wastes could be stored by their generators, if necessary (GAO, 2004). In a recently issued policy position, the LLW Forum stated, "There is not an immediate crisis. The current national waste management system affords flexibility to make adjustments as conditions across the country change; however, it is important to continue working to meet all current and future disposal needs" (LLW Forum, 2005, p. 3).

According to Manifest Information Management System (MIMS) data, the volume of commercially disposed Class B and C wastes has remained nearly constant at just under 900 m³ per year for the past 10 years.[30] States that may lose access to disposal after 2008 produce around two-thirds of this total, or around 600 m³/year. About 90 percent of these Class B and C wastes come from nuclear utilities, while the remainder come from medical and other non-utility sources. By contrast, total Class A waste disposals averaged more than 64,000 m³ per year during the same period but contained well under 1 percent of the curies disposed.

The GAO and LLW Forum report that generators could store their small volumes of Class B and C wastes on-site indefinitely in the worst case. While they acknowledge this in not an optimum solution, they believe that it does not pose a health and safety risk as evidenced by the fact that many of these same utility and non-utility generators store spent nuclear fuel and GTCC sources. Many medical wastes that are highly radioactive have short half-lives and are routinely stored for decay on-site (NRC, 2001a).

Significantly, the GAO received essentially no responses to a questionnaire sent to several thousand radiation control officers asking for their concerns about future access to LLW disposal facilities. This lack of concern seems to mirror the lack of concern when South Carolina left the Southeast Compact in 1995. That action opened the Barnwell site to every state except North Carolina, arguably a violation of the interstate com-

---

[29]The Low-Level Radioactive Waste Forum (LLW Forum) represents the interstate compacts, states that are designated by a compact to host—or that currently host—a commercial LLW disposal facility, and unaffiliated states. Voting members of the Board of Directors are appointed by governors or compact commissions and are authorized to speak for their states and compacts with regard to LLW policy. See http://www.llwforum.org/.

[30]The MIMS database is maintained by DOE to monitor the management of commercial LLW in the United States. See http://mims.apps.em.doe.gov/.

merce clause of the constitution. No generator, including nuclear power utilities, fuel fabrication facilities, or other entities within North Carolina, sought relief.

The committee agrees that the post-2008 situation should be watched closely, but does not perceive an impending crisis. The limited number of disposal options for LLW appears to be an undesirable situation, both in terms of ensured access and market economics. The committee encourages efforts, such as the EPA's ANPR, to expand the disposal options for very large volume, very low activity Class A wastes (see Chapter 4 and Recommendation 1 in Chapter 5).

The committee foresees a number of possibilities that could avoid a post-2008 crisis for Class B and C wastes. First of all, South Carolina might rescind its decision to close Barnwell to states outside the Atlantic Compact. As of 2004, the Barnwell site has approximately 76,500 m³ of its capacity remaining (GAO, 2004).[31] It is also possible that the Andrews, Texas, site may be licensed to begin receiving Classes A, B, and C wastes in the 2008 time frame. While it would be licensed under provisions of the Texas Compact, which includes only Texas and Vermont, the Texas Compact provides a discretionary option for the compact commission to contract for the disposal of waste from outside the compact (LLW Forum, 2005). Lastly, if a generator were in a crisis situation because of lack of access to disposal, the generator could seek relief through USNRC action under 10 CFR Part 62, which provides criteria and procedures for emergency access to nonfederal LLW disposal facilities. All of these contingences indicate that there are options and possibilities for continued access to disposal.

CONCLUSIONS

The committee is not alone in recognizing the need for improving LAW practices. The initiatives described in this chapter—by regulatory authorities, professional and trade organizations, and Congress—consider the actual risks posed by a waste material rather than perpetuating origin-based controls. The described initiatives would

- Increase options for disposing of very slightly contaminated wastes, which comprise the overwhelming volume of LAW;
- Recognize the similarity of wastes that contain naturally occurring uranium- or thorium-series radionuclides; and

---

[31]This would provide approximately 80 years' worth of disposal capacity for B and C wastes at the MIMS-reported disposal rate of around 900 m³ per year.

- Impose consistent federal control of concentrated radioactive materials that were previously not included in the AEA.

Authorities outside of the United States generally base their regulations on the radiation doses that might result from LAW disposal, rather than regulating according to origin. International standards provide dose-based exemption or clearance of very low activity wastes from control as radioactive materials. Several countries have special provisions for disposing of LAW that cannot be cleared from controls, but pose little radiation risk. While there are differences among countries, there is reasonable consistency in their approaches for managing and disposing of LAW, and the categories of waste accepted in surface or near-surface facilities are fairly comparable.

The committee judges that U.S. regulatory agencies and other organizations have made important initiatives toward improving the current system and that there is clearly a need to do so. However, there is no pending crisis in disposal capacity, access to disposal, safety, or any other area that would require a complete near-term overhaul of the system. In the absence of such impetus, there is little will among policy makers to engage in radioactive waste issues, which are certain to encounter opposition—as described in the next chapter.

The committee therefore concluded that initiatives such as those discussed in this chapter are the most realistic options for improvements. Unfortunately, some of these initiatives are faltering. Better coordination and mutual support among agencies and organizations responsible for LAW, including exchange of international expertise, will be necessary to make progress.

3

A Risk-Informed Approach to Low-Activity Waste Practices

Daily life requires constant assessments of risk: Should I drive or fly? Should I eat the fish or the hamburger? Is it safe to cross the street? Should I sell one stock and buy another? How much life insurance should I buy? Sometimes we have some control over accepting a particular risk, but at other times we do not. Generally, we believe that the risks we have some control over are less dangerous or less likely to occur than those imposed on us by others (Slovic, 2000). More formally, risks are averages; that is, they reflect the average likelihood that an event will occur in a population over a given period of time. Because risks are averages, they are of limited value to an individual. Knowing the average risk of death per passenger mile flown per year, for example, might help us decide between flying and driving, but does little to tell us the risk associated with a particular planned flight.

Risk, and decisions about whether to accept or avoid a particular risk, are determined by the probability or likelihood of an adverse result and by its severity. The likelihood of losing a single roll of the dice in a craps game is 51.3 percent; the severity of the consequence of losing depends on the amount of the bet involved. The larger the bet, the greater is the severity of the consequence and the less likely may the player be to accept it. On the other hand, the larger the bet, the greater is the benefit if won. The player's decision will be a value judgment that involves weighing potential benefits against potential losses. Similarly, the severity of the loss may become more acceptable as the probability of winning increases. Like gambling, decisions about the acceptability of health risks from environmental exposures to radiation or other hazards depend on the likelihood of an effect, the severity of the anticipated effect, who is affected, and the potential benefit, if any, afforded by the source of the risk.

There is a large volume of literature on understanding and evaluating risk, managing risk, and communicating risk (e.g., Kaplan and Garrick, 1981; NRC, 1983, 1994, 1996, 2005b; Garrick and Kaplan, 1995; Risk Commission, 1997). Publications of the National Academies have consistently advocated using risk as a basis for policy decisions and have provided guidance for characterizing risk, for making decisions about the best ways to manage risks, and for including a broad range of involved citizens in such decisions. In concert with this viewpoint, the National Academies' Board on Radioactive Waste Management[1] initiated this study because its members believed that the present, mainly origin-based, regulation and management practices for low-activity radioactive wastes (LAW) do not provide a consistent basis for systematically managing their risks: "The current systems for regulating this waste lack overall consistency and, as a consequence, waste streams having similar physical, chemical, and radiological characteristics may be regulated by different authorities and managed in disparate ways" (BRWM, 2002). In addition, this committee noted in its interim report that the public expresses "considerable lack of trust in the LAW regulatory system due to its complexity, inflexibility, and inconsistency" . . . raising "doubts about the system's capability for protecting public health" (NRC, 2003a, p. 5).

During the course of this study the committee came to the conclusion that a "risk-informed" approach would provide the best option for improving LAW regulation and management practices in the United States. A risk-informed approach is based on information provided by science-based risk assessment but includes stakeholders as a central component in decision making. This chapter begins by discussing the concept of risk and concludes with a framework for risk-informed management of LAW.

RISK AND RISK ASSESSMENT

Risk is a common currency that allows regulators and other decision makers to compare different threats to public health from all sources, set priorities among them, choose effective risk reduction strategies, and target those that are most important. Defining the elements of risk involves the "risk triplet," a series of three questions posed by Kaplan and Garrick (1981): What can go wrong? How likely is it to happen? What are the consequences or outcomes?

---

[1]This board merged with the Board on Radiation Effects Research to form the Nuclear and Radiation Studies Board in March 2005.

Risk assessment is the practice of using observations about what is known to make predictions about what is not known about the nature and likelihood of a risk. Risk assessment provides a framework for organizing information in a form that is meant to provide a useful input—both qualitative and quantitative—to risk management decision making. The quality of a risk assessment intended to inform disposal decisions will rest on the quality of the available scientific data; on the extent to which the underlying physical, chemical, and behavioral phenomena are understood; and on how well that understanding and any related uncertainties are reflected in the analysis. Because risk assessment is not a way to scientifically or analytically measure risks, but rather to make predictions about potential risks, characterizing and managing risks necessarily rely on judgment and policy in addition to science.

Figure 3.1 shows the National Research Council's interpretation of the relationship between scientific research, risk assessment, and risk management (NRC, 1983). As the figure indicates, scientific data provide the basis for performing a risk assessment, which in turn provides input to a risk management decision. This framework and the basic risk terminology supporting it have served as the basis for environmental health risk assessment, both regulatory and nonregulatory, since the mid-1980s. The results of a risk assessment are used by regulators and other decision makers, along with information about economics, technological feasibility, politics, and the law, to determine how best to manage a risk.

In the context of protecting public health from environmental exposures, risk assessment involves combining information about hazardous materials of concern (in this case, radioactive materials), the fate and transport of materials in the environment, the exposure of individuals, and the likelihood of adverse health effects associated with these estimated exposures. The result of a risk assessment—the calculated risk—is an estimate of the probability that a particular type of health effect will occur in an exposed population (e.g., 10^{-6}, or one in a million). In practical terms such a result would mean that in a very large population, such as that of the United States, exposed to some harmful agent, an average of no more than one extra person per one million people would be expected to develop an adverse health effect (e.g., cancer or whatever effect was being estimated).

Environmental health risk assessments are generally performed to evaluate risks that cannot be measured. Risks cannot be measured because they are generally quite low. For example, if the risk of excess lifetime cancer from exposure to a given pollutant were estimated to be 10^{-6}, it would not be possible to determine which individuals, if any among the approximately 42 percent of the U.S. population that gets cancer (including treatable cancers such as skin cancer) from all causes, could attribute

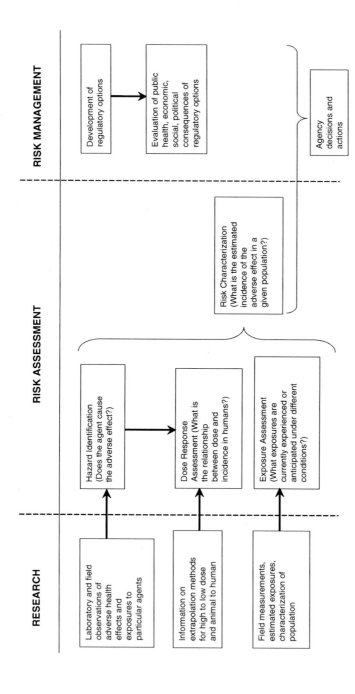

FIGURE 3.1 The National Research Council's interpretation of the relationship between scientific research, risk assessment, and risk management has been a standard for health risk assessment since the mid-1980s. Applied to low-activity radioactive wastes, basic data in the left-hand column would include determining the amount and concentration of radioactive materials. The origin of the wastes is not relevant to determining risk.

SOURCE: NRC (1983).

their disease to that pollutant (NRC, 2005a). In addition, the information used to estimate risks is imperfect and often uncertain for many reasons. For example, the possible migration of radionuclides from a waste disposal facility, one of several steps in estimating disposal facility risks, is difficult to estimate over time. Other factors that add to uncertainty are duration and extent of possible exposures and understanding the likelihood of responses at low doses. Even large populations may not manifest enough health effects to serve as the basis of statistically robust predictions of the health risks from exposures to very low levels of radiation (NRC, 2005a).

Risk from LAW is a function of the probability that an event will occur with release of radioactive materials into the environment, the probability that an individual will be exposed to those radioactive materials, the duration and intensity of exposure to ionizing radiation associated with the released radioactive materials in addition to those already present in natural background, and the probability that the exposure will produce a response. The recently published National Academies' report on health effects from exposures to low levels of ionizing radiation[2] concluded that there is a finite risk of health effects from any exposure to radiation (the linear no-threshold model), but for low exposures ("doses") the probability of inducing a health effect is very low (NRC, 2005a).

Properly done, risk assessment can be a powerful tool for organizing information systematically and understanding the behavior and impacts of LAW under a specified set of conditions and assumptions. Risk assessment of LAW provides a link between the properties of the waste that can be measured or at least estimated (e.g., quantity and concentration of radioactive materials, chemical form, half-life, nonradioactive substrates) and possible adverse effects on human health and the environment (risk), which can only be calculated probabilistically (see Sidebar 3.1; also see Sidebar 3.3 of Appendix A). Risk assessments begin with the measurable properties of a waste material and yield an estimate of the risk associated with its dispositioning (release, storage, disposal) under a given set of conditions. Perhaps more importantly, given the properties of a waste and the level of risk deemed acceptable, a risk assessment can provide guidance about the degree of control needed to achieve that level of risk.

A difficulty in performing risk assessments for waste disposal is developing a comprehensive evaluation of all relevant exposure situations and their associated probabilities during the time period under consider-

---

[2]Specifically, the report dealt with low doses of low-energy transfer radiation, such as X-rays, gamma rays, and low-energy beta particles. The report's conclusions are consistent with previous reports in the Biological Effects of Ionizing Radiation series.

Sidebar 3.1
Some Key Terms Used to Describe Radiological Hazard

Radioactivity

Radioactivity is defined as the spontaneous emission of radiation from the nucleus of an atom and expressed in units of disintegrations per second (becquerels [Bq]). In principle, these disintegrations can be registered on a detector. In practice, measuring low rates of disintegrations in LAW may not be practical because of interference from natural background radiation or because the emissions are shielded by the bulk of nonradioactive materials in the waste. Typically samples of LAW must be specially prepared to allow detection of radionuclides of interest possible.

Radiation Dose

Radiation dose is the term applied to the energy deposited from ionizing radiation an individual receives that is emitted from radioactive material or other sources such as X-rays. Radiation dose is expressed in units of sieverts (Sv) (rem in the United States), with the typical effective dose received by an average individual being about 3.6 mSv (360 mrem) per year from all sources (NCRP, 1987). Doses from LAW can occur to workers handling materials. Doses to members of the public are most often estimated from calculations based on hypothetical events that could breach the isolation system (design failures, intrusion) used to isolate LAW. These calculations can include the probabilities of various events occurring over long time periods commensurate with the radioactive lifetime of the waste. Such calculations are often used in probabilistic risk assessments.

ation, which may be hundreds or thousands of years. To the extent that this can be done, the total risk from all credible processes involving the waste disposal system that may give rise to doses to future individuals can be compared with the level of risk deemed acceptable. Because of the long time periods that must be considered, such a comprehensive evaluation is often not feasible. Hence, as noted in Chapter 2, the International Commission on Radiological Protection (ICRP) suggests a two-pronged approach, where likely or representative scenarios are identified and the calculated doses from these scenarios are compared with the dose constraint. The radiological significance of other less likely scenarios is evaluated from separate consideration of the resultant doses and their probability of occurrence.

Radiation Risk

The relation between radiation dose and health effects has been studied extensively. For the purposes of radiation protection practice, the relationship is assumed to be linear from highest doses that produce acute effects (radiation sickness, death) to lower doses that might disrupt cellular mechanisms and, on a probabilistic or stochastic basis, lead to effects such as cancer. The health risk associated with exposure to low levels of radiation from LAW is calculated using the observed relationship between radiation dose and health effects (principally cancer) at higher doses, using the linear no-threshold model (NRC, 2005a). Such calculations are used for risk-based regulation of radioactive waste.

Factors Affecting Radiation Risk from LAW

Whether a particular LAW poses a health risk is dependent on many factors, including:

- The engineering design and effectiveness of the waste containment or disposal system;
- The fate and transport of the material if released into the environment (e.g., the geophysical and hydrogeological characteristics of the area where it is stored or disposed and the physicochemical nature of the waste and its container);
- Whether a person comes in contact with radioactive material released from the waste (e.g., by inadvertently intruding onto a disposal facility);
- The inherent toxicity of the radionuclide(s) of concern;
- The half-life of the radionuclide(s) of concern;
- The concentration of the radionuclide(s) in the waste; and
- The duration of exposure to radioactive material.

The ICRP view is that this approach does not require precise quantification of the probability of credible but unlikely scenarios occurring in the long term, but rather an appreciation of their probability. Other considerations such as the duration and extent of the calculated doses or risks may be taken into account in determining the significance of such scenarios. The point is that the dose constraint should be interpreted as a comparison with exposure scenarios, and in that context, estimating the likelihood of such scenarios is central. This more qualitative ICRP approach is typical of international risk assessment methodologies.

Based on a particular situation, other aspects of radioactive waste management that could be considered as risks include cultural impacts on resources important to different groups (such as Native Americans);

economic impacts (such as devaluation of property); and psychological damage resulting from fear. Perceived risks can lead to economic impacts and psychological harm even when exposure has not occurred.

RISK AND DECISION MAKING

Risk assessment is only one component of the complex decision-making process referred to as risk management. The risk management process includes consideration of other factors such as economics, technical feasibility, social values or preferences, and legal constraints. Of the various potential approaches to making decisions about how best to manage LAW, there are two that explicitly consider risk: risk-based decision making and risk-informed decision making. This committee endorses a risk-informed, as opposed to risk-based, approach to managing risks associated with LAW, as explained in the remainder of this chapter.

Risk-based decision making relies mainly, if not solely, on the results, most often numerical, of risk assessments. Risk-based decision making is a process of deciding whether, how, and to what extent a risk should be managed based on the magnitude of a quantitative risk estimate (USNRC, 1998). As a hypothetical example of the use of risk estimates, if a risk estimate is 10^{-5} or less, no action may be required; if the risk is between 10^{-5} and 10^{-2}, some action may be needed to reduce the risk; and if the risk exceeds 10^{-2}, more extreme action may be necessary. Risk-based decision making tends to be a prescriptive framework that typically does not permit much interpretation. Considerations of cost, feasibility, special sensitivities, or the relative importance of the risk in a given setting generally are not part of risk-based decision making. Risk-based decisions are generally made by technical experts without benefit of stakeholder involvement or public consultation.

Risk-informed decision making evolved from early risk-based concepts into processes that are more flexible and not guided solely by quantitative risk estimates (NRC, 1983; USNRC, 1998). In the context of this report, risk-informed decision making could involve economic considerations, social concerns, preferences of affected citizens, and other factors in addition to a numerical risk estimate. It might include consideration of risk-risk trade-offs such as extended storage to allow decay over time. Risk-informed decision making acknowledges that risk assessment is more than a mathematical exercise and should be a decision-driven activity, guided by risk management goals and directed toward informing choices and solving problems (NRC, 1996). A more recent study found that the biggest challenges to developing a meaningful risk-informed decision process are minimizing disruption to existing laws, regulations, and agreements and enabling meaningful participation by parties who have few resources (NRC, 2005b).

A general framework for environmental health risk-informed decision making was developed by the Presidential/Congressional Commission on Risk Assessment and Risk Management (Risk Commission, 1997). The Risk Commission was established by Congress through the 1990 Clean Air Act Amendments and was tasked to evaluate and make recommendations about the use of risk assessment and risk management across the federal government. The commissioners were appointed by the President, the majority and minority leaders of the House and Senate, and the National Academy of Sciences.

In its 1997 final report, the Risk Commission concluded that a good risk management decision emerges from a process that elicits the views of those affected by the decision, so that differing technical assessments, public values, knowledge, and perceptions are considered (Risk Commission, 1997). The Risk Commission referred to those affected by a risk or a risk management decision as stakeholders, stating:

> Stakeholders bring to the table important information, knowledge, expertise, and insights for crafting workable solutions. Stakeholders are more likely to accept and implement a risk management decision they have participated in shaping. . . . Stakeholder collaboration is particularly important for risk management because there are many conflicting interpretations about the nature and significance of risks. Collaboration provides opportunities to bridge gaps in understanding, language, values, and perceptions. It facilitates an exchange of information and ideas that is essential for enabling all parties to make informed decisions about reducing risks (Risk Commission, 1997, vol. 1 p. 17).

In the case of LAW, stakeholders could include the waste generators, the agencies responsible for regulating waste disposal, the operators of waste disposal facilities and their workers, and citizens residing near waste facilities and waste transportation routes.

The Risk Commission's framework is shown in Figure 3.2. It is a six-stage process that is circular because decision making should be iterative as new information becomes available. However, the arrow is missing from the last stage to indicate that decision making cannot keep iterating without reaching a conclusion, a process sometimes referred to as "paralysis by analysis." What makes this framework risk-informed is its other salient feature: placing stakeholders in the center of the decision-making process.

The Risk Commission's framework has been used in a number of settings, including the strategy developed by a National Research Council committee for cleaning up polychlorinated biphenyls in the Hudson River

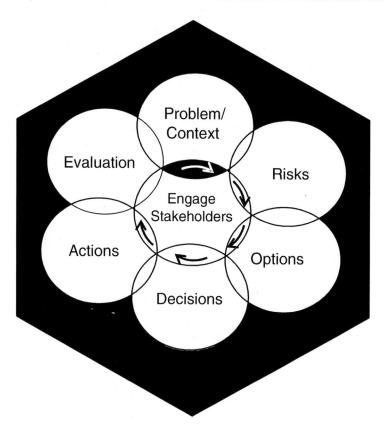

FIGURE 3.2 The Risk Commission's framework for decision making includes six steps. The process is iterative, but it is intended to stop once an objective evaluation concludes that appropriate actions have been taken—for example, when the risk is deemed to be as low as reasonably achievable (ALARA), as discussed later in this chapter. Stakeholders are central to each step of the process.

SOURCE: Risk Commission (1997).

(NRC, 2001b). The U.S. Department of Defense, in partnership with various federal, state, and local agencies, relied on the framework to develop and refine a process for assessing and managing risks from unexploded ordnance. Several states have referred to the framework when they wanted to remind the Environmental Protection Agency (EPA) that it needs to be more open and inclusive of stakeholders in its decisions about air toxins (Kelly Rimer, EPA Office of Air Quality Planning and Standards, personal communication to committee member Gail Charnley). The Canadian Northern Contaminants Program, a multiagency program led by Indian and Northern Affairs Canada, relies on the framework as a tool to help inform northern communities about local contamination (INAC, 2003). The advantage of the Risk Commission's framework is that the risk management goals of stakeholders are established at the outset of the decision-making process and are used to provide a context for risk assessment and to guide choices among risk management options.

RISK PERCEPTION, TRUST, AND STAKEHOLDERS

Different stakeholders have different perceptions of the nature, severity, and likelihood of risk. In particular, technical experts' risk perceptions can vary widely from those of the general public. For radiation issues, risk perception is complex, yet public perceptions about radiation in general and LAW in particular are often important factors in community acceptance of decisions about waste disposal siting and transportation. Most members of the public do not hold consistent perceptions of different types of radiation risks (Slovic, 2000; MacGregor et al., 2002). People see risks from nuclear power or weapons development and production as much higher than risks from X-rays or other medical treatments involving radiation. For example, in a survey of 205 university-age, highly educated adults, a significant number of the respondents believed that exposure to radioactive waste at "even a tiny fraction of current regulatory levels constituted a high or very high risk" (MacGregor et al., 2002, p. 9).

For most citizens, the credibility of and trust in nuclear waste managers is low. In the MacGregor et al. (2002, p. 17) survey, 76 percent of respondents disagreed with the statement that decisions about health effects should be left to experts. Much of the credibility loss for nuclear managers and government regulators has been associated with the Three-Mile Island accident when the public became disillusioned after frequently being assured that nuclear power was safe and that such accidents could not happen (Friedman, 1981), although there were several precursors including the Emergency Core Cooling System hearings in 1972 and the Windscale accident in Great Britain in the 1950s (NRC, 1984). Further losses of credibility occurred in 1988 and thereafter as the media revealed years of mishandled wastes at the nation's nuclear weapons facilities, and accidental and purposeful radiation releases to surrounding areas (Schneider, 1988).

Such negative coverage continues and reverberates, focusing on costly and drawn-out cleanup efforts, further undermining public trust in governmental or private management of nuclear waste disposal (Friedman, 1991, 2001; Ackland, 2002). Today, the impression (real or perceived) that members of the public were not given the complete truth about their exposure to radiation and its concomitant hazards is an important source of fear and mistrust. Communicating with the public about radioactive waste issues has been complicated by the public's lack of trust in those responsible for managing radiation risks and their low credibility.

Risk perception affects trust and the credibility of the risk management decision-making processes. During the early development of risk-based decision making, scientific experts and the organizations they represent dominated risk management. These experts were responsible for estimating risks, and their organizations—often government agencies— were responsible for managing risks that affected the health and safety of individuals. However, as the field matured, the role of experts and technical knowledge in a democracy was frequently debated, particularly in the context of environmental health risk management. The debate centered on conflicts between the "world of values, ethics, politics, and life philosophies" and the "world of information and technical expertise" (Yankelovich, 1991). Scientists were accused of failing to place their efforts in an adequate social context, believing that science is separate from social factors or that social factors play minimal roles (Brown and Mikkelsen, 1990). These differences have been described as technical rationality versus cultural rationality (Krimsky and Plough, 1988).

To include both citizens' concerns and technical knowledge in risk management decisions, decision-making processes involving communities or others affected by risks were increasingly recommended and implemented. For example, the 1996 National Research Council report *Understanding Risk* noted that risk management processes must have an appropriately diverse participation or representation of the spectrum of interested and affected parties, of decision makers, and of specialists in risk analysis at each step (NRC, 1996). The report defined "affected parties" as people, groups, or organizations that may experience benefit or harm as a result of a hazard, of the process leading to risk characterization, or of a decision about risk. The report noted that to be considered affected, such parties need not be aware of the possible harm. "Interested parties" were defined as people, groups, or organizations that decide to become informed about and involved in a risk characterization or decision-making process (who may or may not be affected parties).

A difficult situation that arises often in dealing with environmental issues is when the benefits go to one group and risks are borne by a different group—for example, persons living next to an airport. When members of the public feel they get no tangible benefit personally, they are reluctant to bear any perceived risk. It is perceived benefits versus perceived risks that drive acceptability, and this perception differs among stakeholders. Compensation of some kind is possible in some cases. In South Korea, communities that earlier resisted hosting a LAW disposal site competed for the facility after financial compensation was offered by the government (BBC Monitoring, 2005). Residents of Andrews County, Texas, view the proposed site there in terms of jobs and economic development and are generally supportive of the site (West, 2004).

Despite their common sense appeal, stakeholder-based processes have been criticized for several reasons: the substantial investment of time and resources required; the likelihood that they will heighten, not alleviate, conflict; the difficulty in identifying and facilitating the inclusion of truly representative stakeholders; and the possibility that they are actually counter democratic because of increased involvement of special interest groups (Risk Commission, 1997).

Clashes between the technical and cultural rationalities also draw criticism. Some experts may be concerned that when nontechnical people are included in decision making, the scientific or technical and factual basis of a problem or solution will be distorted, trivialized, or ignored. This problem arises partly because of the difficulty scientists have in communicating technical information as part of stakeholder deliberations and partly because decision makers often perceive nontechnical stakeholders as being more legitimate representatives of social values (EPA, 1995). This clash can also be attributed to nontechnical stakeholders' beliefs that science can be distorted to support different stakeholders' points of view. According to one citizen, "sound science is whatever some expert tells you that supports his or her point of view."

Lack of two important priorities for the sponsoring government agency also can work against successful public and stakeholder participation outcomes and increasing public trust. The first is if the agency initiating the public participation process is not willing (or able) to make the kinds of commitments needed to make the process successful. In such a situation, agency decision makers would not be flexible and open-minded about the nature of the participation and its outcomes. For example, they would not welcome desires of public participants to "redefine problems, focus on different issues, or otherwise change the nature of questions that agencies ask." The second would be if agency decision makers do not recognize the legitimacy of public values and understand that those values may lead to priorities and conclusions that agency personnel, who have their own understanding of what the public interest is, find wrong. According to a recent meta-analysis of 239 cases of public involvement in environmental decision making that had occurred over the past 30 years, a failure to commit to these two important priorities by an agency threatens the legitimacy of the public participation process and whatever public trust the lead agency may have (Beierle and Cayford, 2002, pp. 63-64).

Another problematic factor relates to who should represent "the public" in these processes. Many times, government planners call on a relatively small group of people representing various interests groups to act as "proxy for the larger public." Such smaller groups are needed for a deliberative process, particularly if the issues and goals are complex and require resolving conflicts. However, if the issues involved affect a broad section of the public, then, according to the meta-analysis, broader participation is needed for information sharing and educating the public (Beierle and Cayford, 2002, p. 65). In some complex situations both deliberative and information-sharing activities are required.

Another issue that complicates public and stakeholder involvement is how much influence should be given to the public and stakeholders. Many people agree that the public participation process requires some level of public influence. Yet, in most public meetings, citizens may only provide information and comments and agencies may have little obligation to act on these contributions. According to the meta-analysis, "One of the principal reasons offered for low levels of participant motivation was a perception that the public had little influence over agency decisions." Such beliefs work against building public trust. The analysis showed that "the goal of incorporating public values, which essentially measures the public's influence, is highly and significantly correlated with the goal of public trust. In low-trust situations, then, the public may need to be granted more influence to convince them of the legitimacy of the public participation process" (Beierle and Cayford, 2002, p. 68).

Despite these obstacles, there are many ways that governmental organizations, members of the public and stakeholders can and do work together, ranging from traditional public hearings and public comments procedures to policy dialogues, stakeholder advisory committees, citizen juries, and facilitated mediations. The meta-analysis found that among the important factors leading to a successful outcome was a participatory process that starts early in the discussion of an issue. It also found that the type or process of participation is quite important. For example, public hearings and meetings might be quite useful in improving the quality of decisions, but they are not very effective at either resolving conflicts among competing interests or, more importantly, building trust in institutions. The best participatory process for trust building was found to be negotiations and mediations with advisory, stakeholder, or similar types of committees (Beierle and Cayford, 2002, p. 66).

Although it may seem at first glance to take power away from those with legislated responsibility, by involving stakeholders an agency can better identify serious public concerns and be better prepared to deal with them. While in fact the agency retains all responsibility for the decision, if it ignores consensus recommendations of a stakeholder process then future processes are weakened. The committee recognizes that there are some activists who are completely dedicated to thwarting any change and there is no way they can be accommodated, co-opted, or convinced. Including the appropriate stakeholders to participate in risk-informed decision making is difficult. Still, if a decision is to be made and the agency really listens to stakeholders with the goal of minimizing their irritation and anxiety, there is more of a chance that decisions will go forward and not be mired in controversy.

When risk assessment and risk management are conducted by analysis and open deliberation, scientists and technical experts have an opportunity to interact and work with the public meaningfully by establishing a dialogue about a potential hazard and creating a neutral framework for discussion and collaboration (NRC, 1996). Working together can contribute to increased transparency of the decision-making process, more trust among the involved parties, and some reestablished credibility for managers and government officials who have contributed to the process. Although some involved citizens may remain hostile toward agency decisions, achieving broader acceptance is a reasonable expectation. However, this is not an easy process and it requires genuine respect for public values, careful planning, and a commitment to make the public participation process work from all involved parties.

Sidebar 3.2 provides an example of how improving stakeholder involvement in risk management decisions has helped those decisions to be more reflective of social values and public concerns.

ATTRIBUTES OF A RISK-INFORMED LAW REGULATION AND MANAGEMENT SYSTEM

A risk-informed LAW regulation and management system would combine the principles of risk, risk assessment, and risk-informed decision making and apply them to control LAW according to their actual radiological hazards (see Sidebar 3.1). A waste's potential to cause harm would guide decisions about its regulation, management, and disposal. These decisions would be further informed by views, needs, and constraints from all stakeholders—recognizing that the current system of origin-based regulation and management is deeply embedded in legislation, the current regulatory framework, and commercial infrastructure. The system would operate according to the Risk Commission's risk management framework or a comparable framework: involve stakeholders, agree on risk management goals and on a definition of "acceptable risk," establish a system of LAW classifications consistent with their risks, and identify appropriate disposal options. This section sets out the committee's vision of a fully risk-informed system. The steps toward implementing this vision, given practical constraints and currently available mechanisms, are developed in Chapter 4.

Sidebar 3.2
Stakeholder Involvement Among Native Americans

Native American communities are examples of how stakeholder involvement has led to risk management processes more reflective of social values and public concerns. Until recently, the Native American cultures had little understanding of radioactive material or radioactive waste. Risk assessment and risk communications strategies failed to adequately address the complex issues relating to Native American communities. This failure is of particular concern because Native Americans have been exposed to radiation from fallout during atomic testing, from living adjacent to nuclear waste sites, and from working in the uranium industry as miners, millers, and ore transporters. Yet, policy decisions on how best to manage radioactive waste failed to consider Native American values such as respect, balance, containment, moderation, and reverence. Because Native Americans link their own psychological well-being to environmental stability, the role of psychological healing in decisions about environmental restoration should not be underestimated. More recent efforts to address remediation of radioactive waste-contaminated sites affecting Native Americans have begun to include their perceptions of their environment by including them in the decision-making process.

SOURCES: Dawson et al. (1997); Markstrom and Charley (2001).

Acceptable Risk

At the outset of the committee's study, the National Council on Radiation Protection and Measurements (NCRP, 2002) proposed a very general risk-based system for classifying essentially all hazardous wastes (radioactive, chemical, biological, or mixtures of these wastes) into categories of (1) exempt waste that could be disposed in a landfill or equivalent facility; (2) low-hazard waste suitable for near-surface disposal in an appropriately regulated facility; and (3) high-hazard waste that requires geologic disposal or equivalent confinement. The NCRP's general approach and framework served as useful guidance as the committee focused its attention more narrowly and specifically on the attributes of a risk-informed system for LAW. The committee also found international perspectives and initiatives to be useful in developing its views.

In proposing its risk-based classifications, the NCRP developed the following definitions, which are qualitative but nonetheless useful for LAW (NCRP, 2002):

- *Unacceptable* risks are intolerable. Such risks must be reduced regardless of the cost or other circumstances. The NCRP also considers risks unacceptable if they are "tolerable" but not as low as reasonably achievable (ALARA).
- *Acceptable* risks are below intolerable (i.e., they are tolerable) and they are ALARA.
- *Negligible* risks are so low that further risk reduction using the ALARA principle is not warranted. The NCRP notes that achieving negligible risk is not the goal of ALARA. Acceptable risks that are ALARA might not be negligible.

In a risk-informed system, determining what level of risk is acceptable includes nontechnical risk perceptions along with science-based risk assessments and technical and economic constraints. The risk of a given LAW, manifested by its radiological hazards, would be balanced against the control measures applied—including regulation and physical barriers. Determining the balance point—the acceptable risk—is a public policy decision (see Figure 3.3).

While emphasizing its position that determining acceptable risk is a matter of public policy, the committee recognizes the considerable efforts that have gone into developing semi-quantitative guidelines for acceptable risk, often with considerable public input. For example, the EPA uses an upper bound on acceptable risk over the lifetime of an exposed individual of around 10^{-4} in its Comprehensive Environmental Response, Compensation, and Liability Act cleanup requirements. The ICRP guidelines for "constrained optimization" are equivalent to a risk of about 10^{-5} per year (ICRP, 1998). The committee does not advocate reassessing these guidelines. Rather, these dose-based risks may serve as a starting point for developing risk-informed practices. Risk-informed regulation and practice would then ensure that all LAW is controlled to a consistent level of acceptable risk. Furthermore, establishing a consistent level of acceptable risk would provide a way to harmonize regulation of LAW with that of other hazardous materials as envisioned by the NCRP.

The risk-informed approach would therefore combine information about the risk arising from inherent properties of a given waste and information about technical, economic, and social issues related to a particular disposal approach in a context of what is considered acceptable. Sidebar 3.3 provides a qualitative illustration of this approach and the basic attributes of a risk-informed system for LAW management.

Waste Classification

A waste classification system is necessary for regulating and managing LAW consistently and predictably. For risk-informed LAW regula-

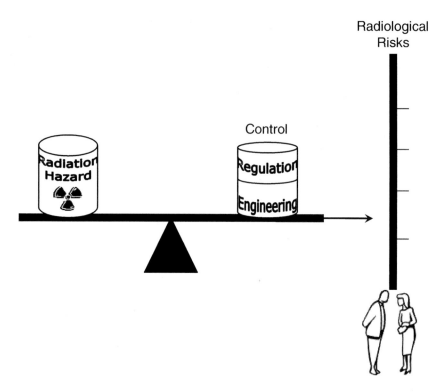

FIGURE 3.3 A risk-informed process for managing low-activity waste will balance the radiological hazards of a given waste with its degree of control by both regulation and physical measures. Wastes that present about the same hazard should receive about the same degree of control. Determining the appropriate balance between radiological hazard and control, hence the risk posed by a management or disposal option, is a matter of public decision making as discussed in Chapter 3.

SOURCE: James Yates, NRSB.

tion and management, the classification system reflects the wastes' actual radiological hazards irrespective of the wastes' origins. Waste classification categories would be developed gradually as the decision-making process evolves away from origin-based to risk-informed. As noted previously, current initiatives in the United States that would lead to more disposal options for truly low-activity wastes and impose greater controls on more hazardous wastes are important initial steps in this evolution. International approaches (e.g., International Atomic Energy Agency and European Commission) to waste classification provide good models for a risk-informed system, as does the overarching system envisioned by the NCRP.

Sidebar 3.3
What Is "Risk-Informed" LAW Management—An Illustration

The objective of waste management is to isolate hazardous wastes away from humans and the environment. No isolation system is perfect, especially for a very long time period, so risks are unavoidable. A useful way to express the concept of risk is to answer the three questions of the risk triplet (Kaplan and Garrick, 1981): What can go wrong; how likely is it; what are the consequences? The triplet can be expressed mathematically, which forms the basis for the science of probabilistic risk assessment. The following qualitative discussion, however, illustrates the basic concepts of a risk-informed approach to LAW management.

Consider a drum of waste placed in a below-surface vault for long-term isolation. One can list what could go wrong: the multilayered cap leaks, the ceiling or walls of the vault crack, someone intrudes, and so forth. Based on experience or other information, one can also assign a probability to each of these events. Thus, at least to a first approximation, the first two questions of the triplet: what can go wrong? How likely is it? are independent of what is in the waste.

The *consequence* of something going wrong depends on the nature of the waste: is it flammable, radioactive, pathogenic, or inert? For LAW, risk is related principally to the radioactive properties of the waste.[a] This is the basis of the committee's argument that the current system for regulating and managing LAW is based on the wrong premise: controlling the waste according to the enterprise that produced it rather than its actual radioactive properties. Some wastes may be overregulated relative to their actual risk; others that pose equal or higher risks may be escaping the regulatory net.

The mathematical methods of risk assessment allow one to estimate a number for the risk associated with a given system for isolating a given waste. Acceptable risk may, for example, be in the range of 10^{-6} to 10^{-4} as discussed in the previous sections. If the estimated risk exceeds that deemed acceptable, then more robust measures to isolate the waste are required.

As discussed previously in this chapter and in other studies, determining the level of risk that is acceptable is a matter of public policy, not purely technical analysis. While not discounting expert analysis, risk-informed decisions explicitly recognize perceptions by all stakeholders, including informed publics. In a facility-siting decision, public officials representing their constituencies may demand a lower risk for a radioactive waste disposal facility than, for example, a chemical plant or a prison, irrespective of the technical assessment. Conversely, decision makers may decide that limited financial resources are better used for reducing risks associated with other activities.

continued

Sidebar 3.3 Continued

Requiring additional measures to isolate the waste also illustrates the principle of ALARA—to be acceptable, radiation doses (hence risk) must be as low as reasonably achievable, taking economic and social factors into account. Even though a risk assessment indicates that a facility meets its risk objective, making improvements that further reduce risk or increase confidence that the facility will work as intended may be appropriate as long as costs are not inordinate.

---

[a]The committee realizes that this illustration oversimplifies an actual risk assessment. A leak into a LAW facility would not necessarily produce a consequence. Even in this simple example, water from the leak would have to rust the drum, interact with the radioactive materials inside, exit the drum, and transport radioactive material out of the vault into the environment. People ("receptors") would have to be exposed to the material and the exposure would have to produce a health effect. Each of these events has an associated probability. Nevertheless, any *consequence* at the end of the event chain is related to the hazardous nature of the waste itself.

The committee did not attempt to develop its own classification system for LAW because deciding what levels of risk may be considered acceptable, unacceptable, or negligible in particular situations is a matter of public policy. Nevertheless, it is possible—and desirable—to describe the attributes of a risk-informed LAW classification system. To be consistent, any classification system should attempt to balance the intrinsic hazards of the wastes with the extent to which they are controlled (see Figure 3.3). Overall, a risk-informed waste classification system would include the following attributes:

1. A lower limit for the radioactivity concentration below which the waste could be exempt from regulation or could be cleared from being regulated as a radioactive material (see Sidebar 3.4).[3] Such wastes present only a negligible risk to health, so they could be disposed of along with other nonhazardous wastes. The committee has not concluded that such exemption or clearance should necessarily imply release into general commerce—rather, conditional release options such as those being explored

---

[3]Presently the only way for wastes to exit the regulatory system is via the little-used case-by-case process under 10 CFR 20.2002, as discussed in Chapter 2.

Sidebar 3.4
Limited Release of Very Low Activity Materials into Commerce

Fly ash from coal combustion typically contains 2 to 10 pCi/gram of NORM, which falls under the regulatory authority of states. In the United States, just over 30 percent of coal ash (14 million metric tons in 2001)[a] is reused and the remainder is landfilled or disposed of in ponds as slurry. Fly ash is being incorporated into cement products, grouting mixes, and asphalt paving. Applications include highway and runway construction, livestock feedlot and hay-storage pads, and a cellular concrete product that can be used as an alternative to wood in floors and ceiling panels. Decisions about whether or not construction products are labeled with information on radioactive content are the responsibility of states.

Under the current origin-based system for regulating LAW, materials with radioactive material concentrations similar to coal ash that are regulated by the USNRC must be disposed in a facility approved for low-level radioactive waste and cannot be released into commerce except on a case-by-case basis.

---

[a]http://www.epa.gov/epaoswer/osw/conserve/c2p2/about/about.htm.

---

by the Nuclear Regulatory Commission's (USNRC's) initiatives on alternative disposition of slightly radioactive solid materials would be available (see Chapter 2).

2. A class of very low hazard radioactive wastes that could be disposed of in appropriately controlled (regulated or permitted) landfills or similar disposal systems. Of particular interest are very low level radioactive wastes from reactor decommissioning and site cleanups considered in the EPA's Advanced Notice of Proposed Rulemaking (see Chapter 2).

3. Classes of wastes that are suitable for near-surface disposal but, due to their radioactive material concentration, total quantity of radioactive materials, and other factors that can affect the risk assessment, require increasingly stringent packaging and disposal methods to meet the agreed-upon acceptable risk objective. The USNRC waste classification system in 10 CFR Part 61 is an example of such waste classes (i.e., Classes A, B, C), and the committee does not advocate changing it. As noted previously in this report, however, the Nuclear Waste Policy Act's definition of low-level waste, which underpins the 10 CFR Part 61 system, is not risk-informed.

4. Classes of wastes that require special consideration for near-surface disposal in order to meet the acceptable risk objective. Portions of nuclear fuel reprocessing waste that may remain at Department of Energy (DOE) sites constitute one example. These wastes are subjects of another National Academies' study (NRC, 2005c). Greater-than-Class C wastes, which will be the subject of a future DOE Environmental Impact Statement, are another example.

5. Classes of wastes that are hazardous to the extent that they require geologic disposal to meet the acceptable risk objective. Most of these wastes have specific definitions under the Atomic Energy Act (AEA; e.g., spent nuclear fuel, high-level waste, transuranic waste). The committee did not discuss these wastes, but includes them here to recognize that they would fit into an overall risk-informed structure (see NRC, 2005b).

It is worth emphasizing that the above attributes of risk-informed categories are independent of the wastes' origins or their current regulatory designations. Regardless of whether they are naturally occurring radioactive materials (NORM) or AEA wastes, some uranium-bearing wastes, for example, might be discharged with little or no control while others would require disposal in a licensed site with engineered barriers—with the degree of control depending on their radiological hazards. As another example, wastes containing about the same amounts and concentrations of ^{60}Co would be subject to the same requirements regardless of whether they arose in a nuclear power reactor (currently AEA controlled) or in a research accelerator (currently under state authorities). As noted above and throughout this report, there are substantial portions of current U.S. legislation and regulation that are consistent with a risk-informed approach to LAW disposal, as are many current or proposed international standards and practices.

CONCLUSIONS

A risk-informed system for regulating LAW—one that regulates radioactive materials based on their radiological hazards rather than their origin—would help simplify and standardize the decision-making process. A simpler, easier-to-understand process will be more open to public scrutiny and participation. The present LAW regulatory system is difficult to understand, even for experts. Because the current system is not easily understood, public participation in decision making is more difficult, which may engender a lack of trust.

A gradual transition from the current regulatory patchwork to a risk-informed system would be an understandable process for all stakeholders and increase their opportunities to participate in the decision making. It could allow people to know which agency is making important LAW decisions, why these decisions are being made, and which agency is accountable if the decisions are wrong. Citizens could be able to better evaluate public health protection and cost-effectiveness to determine whether enhancements are needed. Reducing complexity could also provide more credibility for government officials, showing that they are taking the problem seriously by looking for a more rational and effective way to manage LAW.

4

Implementing Risk-Informed Practices

In this chapter the committee outlines practical approaches to implementing risk-informed practices for low-activity wastes (LAW). The committee found no quick or simple ways to change from the present waste-origin-based patchwork of practices that has evolved over 60 years to the risk-informed system described in Chapter 3. Changes will have to be accomplished within the framework of current laws, regulations, and the large financial investment in management infrastructure, including disposal facilities, that are now in place. The committee also noted the current initiatives from regulators and other organizations as well as the previous lack of success in substantially reforming the system through congressional mandates that are described in Chapter 2.

These considerations led the committee to conclude that gradual, stepwise implementation of risk-informed practices is the best way to proceed. Implementation will require the participation of regulators, waste generators and facility operators, and public stakeholders. Overall, the committee's implementation strategy set forth in this chapter is based on regulatory agencies acting primarily under their existing legal authorities, private industry adopting risk-informed waste practices as good business, and public stakeholders engaging objectively in the decision-making process.

IMPLEMENTING RISK-INFORMED PRACTICES THROUGH THE REGULATORY SYSTEM

The committee in its interim report found that current statutes and regulations for LAW provide adequate authority for protection of workers and the public. In seeking to provide practical advice for implementing risk-informed LAW practices, the committee suggests a four-tiered approach that is within these existing authorities of regulatory agencies, except for targeted congressional action at the highest tier.

Steps in this approach are organized in increasing order of complexity and the time and resources needed to make changes. The committee judges that much can be accomplished by using the simpler approaches first. As knowledge, experience, and comfort that "yes this is the right thing to do" build among all stakeholders, changes may be extended or institutionalized at higher tiers. However, more complex problems may require that solutions originate at the higher tiers. This conclusion can be reached effectively by carefully examining options at the lower tiers.

The committee distinguishes the tiered approach, which is a gradual or stepwise implementation of risk-informed practices, from the current patchwork system of regulation.[1] Every regulatory change under the tiered approach would be made with regard to the inherent hazard of the waste in question and how its hazard compares to the hazards of other waste materials.[2] The tiered system is a means of implementing changes at an appropriate regulatory level and always with a specific direction—a vector—that would eventually lead to a fully risk-informed system as envisioned in Chapter 3. The current origin-based patchwork does not provide this consistency.

A Tiered Approach Toward Risk-Informed LAW Practices

With this perspective, the committee sets forth a four-tiered approach that uses

1.   Changes to licenses and permits of individual waste generators or disposal facility operators seeking solutions for specific wastes, waste streams of a given type, or unique wastes that are infrequently generated.

---

[1]Such a stepwise approach is consistent with other National Academies advice on managing high-level waste and spent nuclear fuels (NRC, 2001c, 2003b).

[2]While the committee included only radiological hazard in its deliberations, other hazardous properties of the waste could be included in the approach being recommended here. The connection between inherent hazard of a material and the risks it poses for handling, storage, and disposal is spelled out in Chapter 3.

2. Changes to guidance documents issued by federal and state agencies that provide interpretations and technical resolutions for specific regulatory issues. This may require the development of Memoranda of Understanding (MOUs) to better align and clarify requirements where there is a shared regulatory responsibility among agencies. Furthermore, agency-specific guidance may be required for management of LAW where one agency has clear and sole authority.

3. Changes to regulations that more formally codify requirements for specific management practices and are promulgated by federal or state agencies under their legislated authority.

4. Legislative changes to basic statutes or definitions that underlie existing laws, regulations, or authorities.

The committee is not recommending the use of any one tier over another. Rather, the committee judges that all of the mechanisms alone, or in combination, provide an available and practical means for implementing risk-informed practices. The balance among the first three approaches is best determined by the agencies that have the authority for regulating LAW. Using case-specific approaches can provide timely and effective solutions, which in turn may pave the way to improved regulations or governing statues. To address an emerging issue, such as a new type of LAW, it might be most effective to begin with a specific license or permit change and move to higher tiers if warranted.

A possible concern with the four-tiered "simplest is best" approach is that changes under existing regulatory authorities might be seen as attempts to avoid public input and scrutiny. The committee notes that the regulatory agencies, and Congress, require public hearings and other opportunities for stakeholder input before changes are made at any level among the four tiers. For each tier, the degree of stakeholder input is commensurate with the magnitude of the proposed change—although provisions for stakeholder input need to be improved as discussed throughout this report. By focusing on a waste's actual hazard and potential risks of a proposed solution for dealing with that waste, rather than on regulatory complexities, risk-informed decision making within the four-tiered approach can increase the ability of public stakeholders to participate effectively.

The Tiered Approach in Practice

This section provides examples of how each tier of the four-tiered approach has been applied to manage and dispose of wastes in ways that appropriately recognize their radiological hazards. The examples demonstrate that implementing risk-informed changes at each tier can be accomplished recognizing the constraints of the current regulatory and management infrastructure. Each tier provides a tool that can be used to implement risk-informed—not origin-based—regulations.

Tier 1: Changes to licenses and permits of individual waste generators or disposal facility operators seeking solutions for specific wastes, waste streams of a given type, or unique wastes that are infrequently generated.

Changes to license and permit conditions are generally considered a part of doing business in the commercial world. The following are two examples where risk considerations have allowed sites to accept additional types of wastes beyond the scope of their original permits or licenses. This flexibility has provided generators with disposal options that otherwise would not have been available and provided opportunities for significant cost reduction.

US Ecology, a subsidiary of American Ecology Corporation, operates a state-permitted landfill near Grand View, Idaho, which was initially permitted to receive only Resource Conservation and Recovery Act (RCRA) Class C waste. A change in this site's permit now allows the site to accept radiologically contaminated waste generated at a U.S. Nuclear Regulatory Commission (USNRC) or Agreement State licensed facility if the material has been specifically exempted from regulation according to a clearly described set of conditions. Wastes that may be accepted include

- Unimportant quantities of source material (i.e., less than 0.05 weight percent of uranium or thorium) uniformly dispersed in soil or other media,
- Naturally occurring radioactive material other than uranium and thorium uniformly dispersed in soil or other media, and
- Accelerator-produced radioactive material.

To be disposed of at the site, wastes in these categories must meet well-defined acceptance criteria based on both concentration and total quantity of specific radionuclides.

This permit change has expanded commercial options available to the Corps of Engineers for disposing of some of its FUSRAP (Formerly Utilized Sites Remedial Action Program) waste, and the site is now a major recipient of these wastes. Since 2001, US Ecology, Idaho, has disposed of about 460,000 m^3 of FUSRAP waste and about 274,000 m^3 of non-FUSRAP NORM (naturally occurring radioactive materials) waste from private-sector clients.[3]

---

[3]Data provided in October 2005 by Simon Bell, US Ecology Idaho, Inc.

Chem Nuclear/Duratek operates the Barnwell, South Carolina, disposal site, which is licensed by the state to receive USNRC Classes A, B, and C low-level waste (LLW).[4] Since the site began commercial operation 1971, there have been a number of license changes that provide exemptions and/or special requirements to allow disposal of additional types of waste. Subject to very specific limitations, the Barnwell site's license conditions now allow disposal of the following:

- Radioactive sources that exceed Class C concentration limits, but with strict limits on total radioactivity and packaging in high-integrity containers (see Sidebar 2.2);
- Radium wastes, including small discrete or diffuse sources, but not including bulk radium-bearing wastes such as uranium tailings;
- Gaseous wastes, including containers with up to 1000 curies of tritium, but internal pressure not exceeding 1.5 atmospheres; and
- Wastes that contain limited quantities of hazardous or toxic materials upon evaluation by the licensee and approval by the state.

These changes were developed through risk-informed negotiation and discussions among the site operator, state regulators, and citizens of Barnwell County, South Carolina. The license changes have allowed the site to provide disposal capability for a variety of wastes for which other disposal options were much more expensive or not available.

Tier 2: Changes to guidance documents issued by federal and state agencies that provide interpretations and technical resolutions for specific regulatory issues. This may require the development of MOUs to better align and clarify requirements where there is a shared regulatory responsibility among agencies.

Regulatory agencies generally provide guidance that can be used to interpret an issue more broadly so that *more than one licensee or permittee* can use the guidance in carrying out their regulated activities. Guidance is often developed jointly by two or more agencies, which serves to demonstrate as well as enhance their cooperation.

Under its authority to establish generally applicable radiation standards to protect the public and the environment, the Environmental Protection Agency (EPA) issues federal guidance for use by federal and state agencies. EPA guidance documents provide principles and policies for radiation protection, and EPA technical reports provide current scientific and technical information for radiation dose and risk assessment. In addition, EPA

---

[4]South Carolina is a USNRC Agreement State.

provides extensive guidance to help generators manage hazardous wastes that also contain radioactive materials (mixed wastes).

The USNRC provides extensive guidance (Regulatory Guides and NUREGs) for licensees seeking to decommission nuclear facilities. Such guidance allows licensees to make technical and business decisions—for example, whether to decontaminate a building or to demolish it, or whether the resulting waste can be left on-site or must be shipped and disposed off-site. USNRC guidance for license terminations has been evolving since the late 1990s. The USNRC and EPA signed an MOU on decommissioning and decontamination of contaminated sites in 2002.[5] Efforts to better risk-inform license terminations have continued with USNRC staff presenting an approach to classify restricted-use sites according to their residual risk using a graded approach.[6]

In addition, existing regulations in 10 CFR 61.58 authorize the USNRC to authorize approaches to waste classification as long as the principle protection requirements of 10 CFR Part 61 are met. Part 10 CFR 61.58 states:

> The Commission may, upon request or on its own initiative, authorize other provisions for the classification and characteristics of waste on a specific basis, if, after evaluation of the specific characteristics of the waste, disposal site, and method of disposal, it finds reasonable assurance of compliance with the performance objectives in subpart C of this part.

This gives the USNRC significant authority and flexibility to use a risk-informed approach to waste classification, while maintaining public health and safety, for LLW or LAW.

In 2000, the Secretary of the Department of Energy (DOE) placed a moratorium on the unrestricted release of volumetrically contaminated material from its site decommissioning and cleanup activities pending a USNRC decision on whether to establish national standards for unrestricted release. Although the moratorium is still in effect, DOE has drafted guidance on control and release of property with residual radioactive contamination (DOE, 2006). This guidance has assisted several sites in the disposition of slightly contaminated sediment and rubble.

Multiagency guidance documents are also important mechanisms for implementing risk-informed practices and for demonstrating cooperation among agencies. The Multi-Agency Radiation Survey and Site Investiga-

---

[5]The MOU between EPA and USNRC on site decommissioning is available at http://www.nrc.gov/reading-rm/doc-collections/news/2002/mou2fin.pdf.

[6]See http://www.nrc.gov/reading-rm/doc-collections/nuregs/staff/sr1757/s1/index.html.

tion Manual (MARSSIM) is a consensus document that was developed collaboratively by DOE, EPA, USNRC and the Department of Defense over a period of approximately 10 years. MARSSIM's objective is to "describe a consistent approach for planning, performing, and assessing building surface and surface soil final status surveys to meet established dose or risk-based release criteria, while at the same time encouraging an effective use of resources."

Tier 3: Changes to regulations that more formally codify requirements for specific management practices and are promulgated by federal or state agencies under their legislated authority.

Examples of regulatory changes that can lead toward risk-informed LAW practices are the EPA's Advance Notice of Proposed Rulemaking for disposing of certain wastes in RCRA Class C landfills and the USNRC's proposed rule on alternative disposition pathways, both of which were discussed in Chapter 2. These proposed rules would increase the disposition options for very low activity wastes in a way that the committee judges to be risk-informed. The committee heard different opinions about the degree to which these agencies are committed to finalizing these rules and their eventual likelihood of success.[7] Nonetheless, the committee endorses these efforts, and encourages the agencies to develop these or similar rules and changes to regulations. They are important attempts for regulatory agencies to use their existing authorities to implement risk-informed regulation.

The EPA and USNRC have cooperated in the development of two other rules promulgated by EPA that the committee views as illustrative of movement toward risk-informed regulation. These rule changes eliminate complex dual-regulation of a variety of low-hazard mixed wastes. In 1998 EPA promulgated its Hazardous Waste Identification Rule for Contaminated Media (40 CFR Part 260). That rule eliminated certain types of mixed LLW from RCRA requirements for storage, transportation, and disposal if they were to be disposed in a USNRC-licensed facility. Subsequently, in 2001, EPA's mixed waste rule (40 CFR Part 266) provided more flexibility to generators and facilities that manage mixed LLW, TENORM (technologically enhanced NORM), and/or accelerator-produced material by exempting them, according to certain conditions, from RCRA Class C requirements. If so exempted, those wastes must be managed as radioactive wastes in accordance with USNRC or Agreement State regulations.

---

[7]The USNRC suspended its proposed rulemaking in mid-2005 due to higher priorities, see Chapter 2.

Tier 4: Legislative changes to basic statutes or definitions that underlie existing laws, regulations, or authorities.

While the committee sees neither the likelihood nor the need for Congress to develop sweeping new LAW legislation, there are clearly instances where specific, targeted legislative actions are helpful and perhaps required. Recent examples cited in Chapter 2 are the 2005 Energy Policy Act's changes to the Atomic Energy Act's (AEA's) definition of byproduct materials. Those changes expand the AEA to provide federal authority to control discrete sources of ^{226}Ra, accelerator-produced radionuclides, and other concentrated NORM sources so designated by the USNRC in consultation with EPA. The Energy Policy Act requires the USNRC and EPA to develop a definition of "discrete" sources—sources with sufficient concentrations of radioactivity to warrant federal control.

In its interim report, the committee noted that these materials were inconsistently controlled by federal and state agencies. By placing the subset of materials with the highest concentrations of activity under AEA control—subject to USNRC licensing—the Energy Policy Act provided a significant legislative step toward risk-informed regulation. The subset of these wastes not designated as discrete sources (sometimes called "diffuse" sources), however, will remain under disparate controls as discussed in Chapter 2.

Previously the committee noted that the definition of "low-level waste" in the Nuclear Waste Policy Act, which is reflected in the AEA, is not risk-informed. The definition is a "catchall" that includes AEA wastes that do not have another statutory definition (e.g., high-level waste, transuranic waste). New legislation would be required to change this basic definition, for example to allow very low activity wastes to exit the regulatory control system as discussed in Chapter 3. The committee judges that enacting such legislation is unlikely, and therefore encourages the Tier 3 initiatives described above and in Chapter 2.

INDUSTRY'S NEEDS AND RESPONSIBILITIES FOR IMPLEMENTING RISK-INFORMED LAW PRACTICES

As the previous section illustrates, a tiered approach can be adopted to improve the regulatory system for LAW. Regulators and public decision makers do not bear the full responsibility for moving toward risk-informed practices—the nuclear industry and other waste generators share this responsibility. This section discusses industry's needs and responsibilities in implementing a risk-informed approach to LAW.

Risk-informed practices are good business practices. By working with regulators, public authorities, and local citizens to implement risk-informed practices, industry can increase the cost-effectiveness of its LAW disposal; increase its options for such disposal; and by moving away from the ad hoc nature of the current origin-based system, increase the predictability of its disposal options.

Contemporary notions of regulatory effectiveness rely on a responsible regulated community working with regulators and involved publics. Under the system of command-and-control regulations that dominated through the mid-1980s, regulators and industry frequently played adversarial roles. A great deal of energy was consumed in challenging the regulatory system, resulting in inefficiencies when regulators specified technology solutions and in unpredictability when courts resolved disputes over compliance dates and other program features. With the general shift to more collaborative and market-sensitive regulatory strategies, the regulated community gains predictability and at the same time shares responsibility for taking actions that protect health and the environment.

For the purpose of this discussion, "industry" includes any institution that takes actions to create, manage, or dispose of LAW that require regulatory approvals. Such an entity may be a for-profit corporation, a not-for-profit organization, or a government body. These entities, however diverse their structures may be, share some common needs in the regulatory process. Cost is of course a vital factor. Too-stringent regulations consume resources that could be employed to greater benefit elsewhere. Industry needs to be part of the process of making regulations, through dialogue with regulators both in the formulation stages and during the comment stages after new regulations are proposed. Organizations such as the Nuclear Energy Institute often provide a good mechanism for such dialogue.

In a recent USNRC workshop on decommissioning, industry officials said they would like to see more flexibility in where to send their waste, consistency in regulations between the USNRC and EPA, and finality of closing a site once the required decommissioning work is complete. Fuel Cycle Facilities Forum Chairman David Culberson stated that "waste disposal is typically the largest single cost component of decommissioning, and frequently licensees are left with only one commercial disposal alternative. The industry would like to see more facilities available for disposal" (D&D Monitor, 2005, p. 6).

Industries that generate LAW can take a series of actions that will facilitate and accelerate the transition to a risk-informed system. These actions are consistent with norms of institutional responsibility and based on general principles of transparency, accountability, and sustainability. They include the following:

- The polluter pays,
- Minimize waste,
- Report to stakeholders,
- Invest in worker training and protection,
- Share best practices and lessons learned—negative as well as positive,
- Act in advance of regulatory requirements,
- Develop a sustainability strategy.

The Polluter Pays

In most countries, substantial efforts are taken to ensure that the costs of waste disposal are borne by the entities that produce the waste and are not subsidized by governments (citizens). This is referred to as the "polluter pays principle." The increasing costs of sound disposal serve as a large incentive for organizations to make waste reduction efforts.

Within U.S. corporations, steps have been taken to remove waste disposal from general overhead accounts and to allocate disposal costs to the business unit that produced the waste. Full cost accounting (sometimes called environmental full cost accounting or total cost assessment) is a management tool used by some companies to quantify the costs and then allocate them in a way that will best achieve the objective of reducing waste. This internal accounting strategy has been associated with substantial pollution prevention activity (Rondinelli and Berry, 2000).

The costs of the DOE's waste programs are borne by taxpayers. Nonetheless, public oversight of DOE's budget provides incentives for cost-efficient LAW practices (GAO, 1999).

Minimize Waste

The polluter pays principle is credited with a range of actions that reduce the volume of waste. In the United States, the Pollution Prevention Act of 1990 identifies four categories of waste reduction:

1. Equipment, technology, process, or procedure modifications;
2. Reformulation or redesign of products;
3. Substitution of raw materials; and
4. Improvement in management, training, inventory control, materials handling, or other general operational phases of industrial facilities.

There are many examples in which companies have made dramatic reductions in hazardous waste in each of the above categories, with benefits to worker health and the environment, and cost reductions to the generator. For example, 3M reports that in the last 30 years (1975-2005), projects of this type have prevented 2.2 billion pounds of pollutants and saved the company nearly one billion dollars.[8] Considerable effort has been invested in the development of decision-making tools, both technical (OTA, 1992; EPA, 1994, 1997) and financial (EPA, 1992), to facilitate pollution prevention practices.

Waste minimization (practices to reduce the amounts of waste generated) and waste segregation (avoiding the mixing of less hazardous waste with more hazardous waste) have long been cornerstones of radioactive waste management (IAEA, 1987). The Government Accountability Office (GAO) report on waste disposal capacity in the United States (GAO, 2004) noted that waste generators are making concerted efforts to reduce waste volumes. Such techniques include substituting nonradioactive materials for radioactive materials, keeping nonradioactive wastes free from radioactive contamination, internal recycling, compaction, and incineration. According to GAO, some USNRC licensees have supercompacted Class A wastes to achieve up to a 500-fold reduction in volume or reduced combustible waste to ash through incineration.

France, which like the United States does not allow free release (clearance) of very low activity waste from regulatory control, enforces strict zoning in nuclear facilities. The practice requires identifying "conventional waste zones," where there are no radioactive materials, and "nuclear waste zones" in and around the facilities. Zoning helps ensure that conventional wastes are not contaminated by radioactive materials. Primary controls to ensure this segregation include the facility's design, its operating procedures, and its history—including design modifications and operating incidents. Secondary controls include instrumentation to detect radioactivity in wastes exiting the conventional zones. Optionally, the conventional wastes may be measured again upon arrival at a disposal site for nonradioactive materials (Averous, 2003).

Report to Stakeholders

In the interest of transparency, many institutions make available extensive data on resource use and waste, often in the form of an environmental or sustainability report. Governments may require reporting of select industry activities related to the environment, however; many company stakeholders have expressed interest in having more information than required by government.

Providing data that allow comparisons within industry sectors and from one sector to the next is an important element of transparency to

---

[8]See "Pollution Prevention Pays" at http://solutions.3m.com/wps/portal/_l/en_US/_s.155/113842/_s.155/115848.

many interested stakeholders. For example, the socially responsible investment community makes a range of data-driven decisions, and some investment firms select companies for their portfolios that are considered "best in class" using comparative data. In addition to a variety of external stakeholders with an interest in these data, the expectation is that responsible companies will use the data internally as they follow the adage of "managing what they measure" and reducing the risk associated with the waste they produce.

Increasingly, institutions are implementing comprehensive environmental management systems such the Eco-Management and Audit System now required in the European Union and Japan, or ISO (International Organization for Standardization) 14000, a similar generic environmental management system. These systems have several elements, including commitments to continuing progress in pollution prevention and to reporting. Increasingly there is interest in establishing comprehensive environmental performance measurement schemes such as the Global Reporting Initiative, which will allow quantitative comparisons within sectors and across organizations; pollution prevention efforts are at the center of measurement schemes as well. Characterizing and reporting on LAW is consistent with several trends in corporate responsibility and can be accomplished without government regulatory action.

Invest in Worker Training and Protection

Industries that use materials or processes that are potentially hazardous to health and safety make significant investments in worker training and protection to safeguard their human resources and reduce lost time associated with accidents. Responsible industries invest in more comprehensive worker training; instead of simply protecting against losses, expanding training can result in savings. Many of the waste minimization savings cited above are the result of personnel, often working in teams, trained to identify pollution prevention opportunities and to experiment with process modifications. In a risk-informed system for managing LAW, responsible institutions will examine their worker training protocols to determine whether they are sufficient both to protect against losses and to increase the likelihood of generating savings.

Share Best Practices

Sharing best practices among organizations in the same industry sector and across sectors is an approach that is used to accelerate the rate of technology transfer and to address stakeholder concerns. Global organizations such as the World Business Council for Sustainable Development and the World Association of Nuclear Operators emphasize sharing knowledge among member institutions and, through their publications, share information with a wide range of other organizations. Programs within industry sectors, such as the chemical and nuclear industries, emphasize sharing best practices for the additional reason that an incident involving a single facility may have negative repercussions for the industry as a whole. Industries generating LAW will benefit from sharing a range of best practices, including waste characterization and waste minimization techniques.

Act in Advance of Regulatory Requirements

Institutions have articulated several reasons for acting in advance of regulatory action, including

- Enhanced reputation among customers and other stakeholders;
- Greater control over timing of action, particularly when capital investments are required; and
- Increased likelihood that when regulations are implemented, regulators will codify elements of successful practice, conferring a competitive advantage on the organizations already following those practices.

Many companies using and producing ozone-depleting chemicals acted in advance of the deadlines established for substance phaseout in the Montreal Protocol. For example, several companies that used ozone-depleting chlorofluorocarbons (CFCs) for cleaning found that substitute processes that used water or abrasive beads were less costly than the chemicals they had been using. DuPont, a major producer of CFCs, also had invested in research on alternatives. These factors, along with considerable public concern about the predicted effects of ozone layer depletion, contributed to a situation in which acting in advance of regulatory requirements was a prudent and responsible course of action for many companies.

In a risk-informed system for managing LAW, there will be changes in the degree of regulatory oversight experienced by different industry sectors. For those sectors in which the current patchwork system creates gaps that will be filled by regulatory action in a risk-informed system, acting in advance of government requirements may be a valuable strategy for generators of LAW.

Develop a Sustainability Strategy

Increasingly, institutions are taking concerns for sustainability into account as they develop their long-range strategies. The specifics will vary dramatically depending on the nature of the industry, ranging from changes in the business model to making a transition from one business to another. Two companies that are associated with changes to their business model are Interface Carpet and Xerox.

In response to concerns about the amount of carpeting disposed in landfills, Interface took a number of actions, including development of a product that used recycled fiber, and began to offer carpet in the form of tiles so that only the worn portions could be replaced. Following these actions, Interface began experimenting with the idea of renting carpet rather than selling it to consumers. The rental concept is a form of product stewardship, ensuring the company a stock of used carpet to recycle.

Along similar lines, Xerox several years ago began to increase the recycled material content of its copying machines. The goal was to minimize the amount of material ultimately disposed to land. In the first year of its program the company saved more than $50 million by modifying logistics, inventory, and material purchases (Murray and Vietor, 1995).

Companies generating large volumes of LAW may use sustainability considerations as motivators for development of new business models or as the rationale for a transition from one business to another. Reasons for considering sustainability as a critical element in long-term strategy include the reasons cited above for acting in advance of government regulation along with an added element: product differentiation.

Conclusions About Industry's Needs and Responsibilities

Responsible actions by industry can accelerate the implementation of a risk-informed system for LAW, particularly if institutions act in advance of regulatory action. As the examples offered in this section indicate, there are cases in which responsible actions by industry have resulted in tangible benefits, including

- Enhanced reputation among customers;
- Reduced costs of operations; and
- Reduced costs of disposal.

The committee judges that risk-informed practices are good business practices. By working with regulators, public authorities, and local citizens to implement risk-informed practices, industry can increase the cost-effectiveness of its LAW disposal; increase its options for such disposal; and by moving away from the ad hoc nature of the current origin-based system, increase the predictability of its disposal options. Given the degree of public concern associated with radioactive waste, institutions taking responsible actions related to their LAW also may experience an enhanced public image.

PUBLIC STAKEHOLDER INTERACTIONS IN IMPLEMENTING A RISK-INFORMED DECISION-MAKING PROCESS

Implementing a risk-informed decision process that involves significant public stakeholder participation is not an easy task for several reasons. First, the issues associated with LAW are complex and controversial. Most members of the public see radiation risks from a narrow frame of fear and danger, as described in Chapter 3. Second, despite the efforts of various government agencies to include public stakeholders in decision making, many members of the public are distrustful of these agencies. Consequently, establishing a truly effective public participation process is often difficult. Third, many members of the public and public stakeholder groups have a difficult time understanding the strange and convoluted regulatory system this nation uses for LAW. It does not engender public trust and confidence. In parallel with the previous section's discussion that risk-informed practices are good business practices, this section develops the thesis that risk-informed practices provide effective new opportunities for involving stakeholders and reducing mistrust.

Currently there are a number of LAW issues that worry public stakeholders, and any action that appears to lessen regulatory control of low-level radioactive wastes is a major one. In a March 2005 petition, representatives of Public Citizen, the Sierra Club, Physicians for Social Responsibility, Friends of the Earth, Greenpeace, and other organizations and individuals urged the USNRC to hold a public meeting on the proposed rulemaking for alternative dispositions of slightly radioactive wastes (the proposal is discussed in Chapter 2). The petition sought an opportunity for representatives from their stakeholder community to testify to the commissioners about why the USNRC should not proceed with its staff's proposal to "deregulate significant portions of the 'low-level' radioactive waste stream, permitting licensed nuclear material to go to unlicensed sites such as local municipal garbage dumps, hazardous waste sites, and recyclers for use in consumer goods and construction material" (Public Citizen, 2005).

One part of this very controversial issue is what to do with very low level radioactive wastes from decommissioned nuclear plants. Placing such wastes into a landfill has been done successfully with community involvement and negotiation at Big Rock Point in Michigan, but there are citizens in California and elsewhere who oppose such a possibility (Lucas, 2002).

Citizen concerns with the USNRC's proposed rulemaking evidently were heard and played a role in its decision to defer further action. In voting to disapprove the rulemaking in June 2005, USNRC Chairman Diaz noted that there had been multiple public workshops and public meetings to obtain a range of stakeholders' views. He said the decision-making process for the proposed rule was open and reflected "extensive stakeholder input from citizens and environmental groups, metals and concrete industries, nuclear industry representatives and other state and federal agencies, representing a broad-based and diverse set of views." He called this process "enhanced participatory rulemaking" (Diaz, 2005b).

Another concern relates to the exclusion of interested publics, but not affected publics, from the siting process of an LLW repository in Texas.[9] According to testimony by Dr. Melanie Barnes (2003), representing the League of Women Voters of Texas: "There has been no public discussion by the citizens of Texas about the disposal of federal low-level radioactive waste in Texas. In fact, the current wording of these bills[10] diminishes the right of the public to be involved or even informed by restricting public involvement to those residing in the county where the proposed waste disposal facility would be located. Why should a few citizens be allowed [to] make such a complex and long-lasting commitment for all the citizens of Texas?"

For technical professionals, it is very sensible to regulate LAW based on the hazards posed by type and level of activity rather than by generator. Further, it is logical to use a single set of radiation protection criteria to protect people in all states and countries. The committee nevertheless recognizes the difficulties that will arise if changes in the current system appear to reduce regulatory controls. Citizens of a state with stricter standards may be unwilling to accept less stringent federal statutes. Similarly, on a larger scale, the U.S. public may be unwilling to accept less stringent international statutes of the European Union or the International Atomic Energy Agency. People who are aware of the linear, no-threshold model (NRC, 2005a) may be less likely to accept rules that might increase their potential exposure, even if officials in Washington or Vienna assure them that the risks are insignificant.

These concerns are of major importance politically and socially. They are part of the cultural rationale that heavily influences how stakeholders respond to governmental and nuclear industry actions regarding LAW.

---

[9]NRC (1996) defined "affected parties" as people, groups, or organizations that may experience benefit or harm as a result of a hazard, or of the process leading to risk characterization, or of a decision about risk. "Interested parties" were defined as people, groups, or organizations that decide to become informed about and involved in a risk characterization or decision-making process (and who may or may not be affected parties). See Chapter 3.

[10]SB824 and HB 1567.

As noted in the Risk Commission report discussed in Chapter 3, stakeholders need to be involved early and to be central in any decision-making process. Developing a risk-informed approach to decisions about LAW will ensure that stakeholders are involved early and remain at the center of the decision-making process.

Currently, there are a number of public outreach programs in the federal government to gather information about nuclear issues from, and provide it to, public stakeholders, although not all of them put stakeholders in a central decision-making role. According to the USNRC (2002, p. 1), the commission "has long recognized the importance and value of public communication and involvement as a key cornerstone of strong, fair regulation of the nuclear industry." The USNRC has developed three categories for public meetings to inform citizens and listen to public comment.

In Category 1, the most limited category, the public can observe the USNRC's interactions with licensees and offer constructive comments. In Category 2, public interest and citizens groups can participate in meetings with groups of industry representatives, licensees, and vendors, providing opportunities for the public not only to observe and obtain factual information but also to provide feedback on issues, alternatives, and decisions. Category 3 meetings "provide an opportunity for NRC and the public to work directly together to ensure that the issues and concerns are understood and considered by USNRC" (USNRC, 2002, p. 6). Examples of the latter types of meetings are town halls or roundtable discussions, environmental impact statement scoping meetings, and proposed rulemaking meetings.

A number of other examples of public outreach concerning nuclear issues exist. The Hanford Thyroid Disease Study conducted by the Fred Hutchinson Cancer Research Center and sponsored by the U.S. Centers for Disease Control and Prevention had a public advisory board and also a health information network to keep the public informed by means of a website and newsletters during the nine-year study (Friedman, 2001).

EPA in its recertification program for the Waste Isolation Pilot Program (WIPP) in New Mexico has many fact sheets for the public on a special website as well as a special WIPP information telephone line and a listserv. On its website, it has a quick guide to public outreach activities, both what it is doing to let the public know what is happening and what the public can do to provide information and feedback to EPA (EPA, 2005a,b).

To prepare for the recertification process, EPA commissioned an outside evaluation of its earlier WIPP public certification outreach program to identify its strengths and weaknesses and lessons learned. Among many findings were that "EPA far exceeded regulatory requirements for public outreach and performed many of the elements of its public outreach program extremely well. . . . However, the restrictions placed on EPA as part of its regulatory rulemaking process prevented it from fully achieving its stated commitments under its public outreach program. . . . EPA was unable to involve the public in the key aspects of the decision-making in which they were most interested" (Phoenix Environmental, 2001).

During the WIPP recertification process, EPA has been working with stakeholders to address their concerns. For example, Citizens for Alternatives to Radioactive Dumping and the Loretto Community believed that the geologic characterization of the subsurface surrounding the WIPP repository did not adequately identify the presence of karst.[11] As a result of these concerns, EPA agreed to reevaluate the potential for the presence of karst at WIPP and the possible impacts of the long-term containment of waste for WIPP (EPA, 2005a).

Some countries in Europe have also enabled stakeholders to participate in radiation decision-making procedures. In France in the late 1990s, when results from epidemiological studies raised causal questions about cases of leukemia in residents living in the vicinity of La Hague, a nuclear reprocessing plant, a group of stakeholders, including members of the public, were brought together to review the situation. In Sweden, a site investigation for developing a high-level nuclear waste repository has carefully taken into account local stakeholder concerns and positions. At one potential site in Oskarshamn, already the home of a nuclear power plant, Mayor Torsten Carlsson said that there had been an open and participatory process with the community and SKB, a company owned by nuclear power plant operators that is responsible for investigating potential repository sites. Final site approval will be in the hands of both the national government and the municipality.

Public engagement has been important and SKB representatives have come to meetings of local clubs and groups. Citizen and environmental group participation has been encouraged, and high school students have even been asked their opinions on nuclear waste issues. Said the mayor: "We [citizens] build competence so we can ask the difficult questions. We ask until we get clear answers. We make them [SKB] translate their technical reports" (Carlsson, 2002). Formalizing this requirement in its document granting permission for a site investigation, the Municipality of Oskarshamn noted that "the authorities as well as SKB must deepen their dialogue with the citizens in order that their issues should guide

---

[11]Karst is a type of geology in which there are numerous sinkholes and large voids such as caves.

regulations and safety analyses. These areas must not be reserved for the authorities' and SKB's experts" (Oskarshamn, 2003, p. 1).

The European Community also supports projects aimed at improving stakeholder involvement in decisions or improving their expertise concerning radioactive risks. One of these projects, Community Waste Management (COWAM),[12] has led to a European network of local stakeholders concerned about radioactive waste management. The initial objective of COWAM was "to contribute to the improvement of the quality of decision making at the local level in nuclear waste management. The purpose was not to determine which technical option is the best for a particular type of waste, but to discuss the quality of the decision-making process from the local level viewpoint." In creating a network of local people from communities involved in nuclear waste management, COWAM wanted to contribute to the people's empowerment. Network members exchange information about their experiences to analyze local involvements in national decision processes and issue recommendations for improving those processes.

Besides outreach, another way in which a few government organizations in Europe and the United States have helped public stakeholders become more central in risk decision-making processes is by helping them hire their own technical experts. For example, experts from the United Kingdom and other countries traveled to Oskarshamn, Sweden, to advise and review issues with the municipality and with citizens (Carlsson, 2002). In the United States, the Office of the U.S. Nuclear Waste Negotiator, which operated under congressional authority as an independent agency from 1990 to 1995 to solicit volunteer localities to host temporary or permanent commercial spent fuel sites, made grants of up to $100,000 to public groups to hire experts and perform self-directed studies of the science and potential risks related to such facilities.

Despite these public-centered activities, at least in the United States there is still skepticism among citizens about how seriously their concerns are regarded by government and nuclear industry officials—that is, how central they really are to the decision-making process. Some people consider many public meetings and outreach efforts "sham procedures," with public concerns about radiation doses, among other things, not taken seriously. While public meetings are often good for sharing information and hearing public comment, they often are not the most effective method for involving citizens and stakeholders in a decision-making process, according to a meta-analysis of 239 public participation case studies (Beierle and Cayford, 2002). Citizens' frustration at not having their

---

[12]See http://www.cowam.com/rubrique.php3?id_rubrique=12.

opinions count in the decision-making process often leads to lowered levels of public trust, as discussed in Chapter 3. Making efforts to limit citizen participation by rules, such as in the case of the Texas League of Women Voters, only adds to the loss of public trust. While a risk-informed system might not solve all the problems involved in LAW public participation efforts, offering public stakeholders' a central role early in the decision-making process should increase trust and add important dimensions to public input into radiation decision making.

A risk-informed system by its very definition would ensure that more information than just scientific risk assessment would be used when evaluating LAW issues. It should give members of the public more confidence that their voices will be heard in this process. However, to truly engage the public and enable its trust, a risk-informed system would have to show that inputs from public stakeholders would be a part of the decision-making process and be taken into account *seriously* despite potential problems with technical versus cultural rationalities. To enable such participation, efforts beyond public meetings and hearings would be needed, including public advisory committees, citizen juries, negotiations, and facilitated mediation. These more intensive public participation efforts would help put stakeholders into the central role envisioned in the risk-informed process.

Another important factor would be to have an even playing field among various stakeholders so that citizen groups have as much visibility and influence as lobbyists. Facilitated discussions among public stakeholders and government officials would point out differing viewpoints related to cost-benefit concerns, safety, credibility of the site management, and degrees of trust in the overall regulatory system. While perhaps time-consuming, discussions on these important issues could bring more transparency to, and trust in, the LAW decision-making process.

Because a risk-informed system is simpler than the patchwork system currently in place, it would allow for better public understanding of the whole LAW process, not just the disposal end of it. People would be able to understand which activities generate LAW, why the waste is generated, and the benefits the public derives (if any) from these activities. It also would make it easier to explain how this process places a high priority on public safety.

Being able to understand a simplified risk-informed LAW process could also allow public stakeholders to evaluate for themselves other options for LAW disposal and the relative costs of each. For example, having such an ability would be important for the question of whether slightly contaminated metals might be allowed to reenter commerce (Chen and Moeller, 2003). Being aware of and understanding the risks involved with "release, recycle, refabrication and reuse in a host of unlabeled consumer products" could help individuals better evaluate safety concerns (Johnsrud, 2005).

If a risk-informed process is to work effectively, members of both the affected and the interested publics must play active and involved roles. They have a responsibility to become and remain informed about the complex issues involved in LAW management by going to town meetings, reading fact sheets, and reviewing websites. For example, for WIPP recertification activities, EPA recommends that members of the public regularly check its WIPP website and enroll for e-mail alerts about meeting announcements, new information, and other news. On the WIPP website, EPA has a section called "Radiation in the News" with links to articles that citizens could use to stay informed not only about WIPP but also about many other nuclear issues (EPA, 2005a).

Besides becoming informed, citizens need to communicate their concerns from the beginning of the process to government and nuclear industry personnel. EPA and other government agencies invite calls, e-mails, or faxes to staff members with any questions and comments and urge that the public be aware of the time periods available for public comments. People can also provide input at public meetings or invite government officials to attend meetings of local clubs or groups. Throughout the risk-informed decision-making process, public stakeholders should keep clear lines of communication open and actively serve as constructive partners.

Local governments, too, have significant responsibilities to their constituents to make sure that both they and their citizens play a central role in a risk-informed process, as was done in Oskarshamn, Sweden. Too often, organized national or state environmental or citizen groups are the only voices representing public stakeholders in nuclear waste issues. While these national groups make valuable contributions to the risk "conversation," local voices also need to be heard and encouraged. Both interested and affected publics need to be active and central partners in a risk-informed decision process.

CONCLUSIONS

Government agencies, nuclear waste managers, and public stakeholders all play an important role in effectively implementing risk-informed processes for managing and regulating LAW. As outlined in this chapter, such a process would offer advantages to each participant in the risk conversation and make it easier for participants to have a less acrimonious, even fruitful partnership. Because of its stakeholder centrality, a risk-informed decision process could help create more public confidence not only in the process but also in the government agency and waste management personnel making radiation risk decisions. Because the process is transparent, understandable, and guarantees that a variety of views will be heard and evaluated, implementing a risk-informed decision process would be an important step in building public trust in the LAW regulatory process.

5

Findings and Recommendations

Throughout its information-gathering activities, the committee received a clear message: A more consistent, simpler, performance-based and risk-informed approach to regulating and managing low-activity wastes (LAW) in the United States is needed. The committee heard nearly unanimous views that a complete conversion of the present origin-based patchwork of regulations and practices to a coherent risk-informed system would be the most desirable way to improve the system. The same presenters, however, cautioned that such a conversion—for example, through congressional action—would be virtually impossible given the long history and investment in the regulatory and operational infrastructure of the current system, the disruption that a sweeping change could cause, and the lack of political will to effect such a change.

The committee found no easy solutions. Optimistically, however, Congress, the Nuclear Regulatory Commission (USNRC), the Environmental Protection Agency (EPA), and other organizations are already developing initiatives that, viewed collectively, move toward a more risk-informed system. The initiatives, which were described in Chapter 2, would increase the flexibility and number of disposal options for very low activity wastes while imposing or maintaining consistent controls on more hazardous and concentrated materials.

The committee concluded that while there are no easy solutions, it is possible to move in incremental steps toward a more risk-informed system for controlling management and disposition of radioactive materials. In contrast to the patchwork evolution of the past 60 years, stepwise implementation would move in a constant direction: away from regulating

LAW according to how or when it was generated and toward regulation based on the actual hazard of the material. Risk, as perceived by generators, regulators, concerned citizens, and elected officials, can provide a common basis—a common currency—leading to better cooperation, agreement, and progress for all stakeholders.

Recommendation 1

The committee recommends that low-activity waste regulators implement risk-informed regulation of LAW through integrated strategies[1] developed by the regulatory agencies. Improving the system will require continued integration and coordination among regulatory agencies including the USNRC, EPA, the Department of Energy (DOE), the Department of Defense (DOD), and other federal and state agencies.

While current statutes and regulations for LAW provide adequate authority for protection of workers and the public, current practices are complex, inconsistent, and not based on a systematic consideration of risks. More efficient and uniformly protective management of the risks posed by these wastes will require moving away from the present origin-based regulatory system—a system that is firmly established through decades of practice and involves a number of federal and state agencies that have different authorities.

The concepts of a risk-informed system developed in Chapter 3 of this report and the implementation approaches described in Chapter 4 would provide the basis for the strategies, which should incorporate the tiered approach set forth in Recommendation 2. The strategies would include legal, regulatory, and implementation issues at a level of detail greater than could be attempted by this committee.

The development and use of integrated strategies would strengthen waste regulators' ongoing efforts to improve LAW regulation and management practices by

1. Focusing the attention of decision makers at all levels on the needs for and benefits of implementing risk-informed practices,

2. Providing a unified approach to developing risk-informed practices that is recognized by all stakeholders as cooperative and mutually supportive, and

---

[1] By "integrated strategies" the committee means the results of agencies working together to develop a single or joint strategies for using the approach in Recommendation 2 to implement risk-informed practices. Because the regulatory agencies have different legal authorities they may develop separate, but integrated, strategies.

3. Promoting harmonization (consistency on the basis of risk) in changes at each of the four tiers discussed in this report.

An important purpose of interagency strategies would be to help regulatory agencies balance their use of the four-tiered approach, including instances where targeted legislation[2] might be needed if the first three tiers are not sufficient for developing solutions.

Chapter 2 of this report described initiatives by the EPA, USNRC, and other organizations that are important steps toward moving from origin-based to risk-informed regulation and management practices. However, some of these individual initiatives are faltering.[3] On the other hand, cooperative interagency efforts have made significant progress in improving regulations in areas that are relevant to LAW management and disposal. Examples include development of the Multi-Agency Radiation Survey and Site Investigation Manual (MARSSIM)[4] and guidance from the Interagency Steering Committee on Radiation Standards (ISCORS), the latter of which includes eight federal agencies and has the goal of improving consistency in federal radiation protection programs. Development of the integrated strategies should build on the successes of MARSSIM, ISCORS, and similar interagency efforts and make even greater use of such efforts.

While it is beyond the committee's task to prescribe how regulatory agencies should do their work, the committee judges that coordinated leadership by federal agencies will be essential—for example, by following the model of a federal committee such as ISCORS or a similar interagency group to further identify and prioritize risk-informed improvements in regulating LAW. Developing and instituting implementation strategies may require several years, as did the work on MARSSIM.

Two areas identified in this study exemplify where risk-informed regulations would improve the current system, and could provide a focus for development of the strategies:

• Wastes containing uranium or thorium and their radioactive progeny generated by Atomic Energy Act (AEA) and non-AEA-controlled industries pose similar hazards (according to the type and concentration

---

[2]The 2005 Energy Policy Act's expanded definition of byproduct materials is an example of such legislation. See Chapter 2.

[3]The USNRC put its proposed rule "Radiological Criteria for Controlling the Disposition of Solid Materials" on hold due to higher priorities. EPA is considering how to proceed after receiving some 1500 comments on its advanced notice of proposed rulemaking for disposing of certain LAW in landfills—one of the alternatives in the USNRC initiative.

[4]See Chapter 4.

of their radioactivity) but are controlled under very different regulatory regimes.

- There is no generalized provision for wastes that contain very low concentrations of radioactivity to exit the regulatory system, although there are examples of case-by-case exemption or clearance of some such wastes.

The many sectors of the U.S. economy that produce or manage LAW will necessitate that USNRC, EPA, DOE, DOD, other federal agencies, state agencies and their representatives, and citizens and citizens' groups have roles in developing the strategies. The committee is aware that federal and state agencies each have mechanisms for obtaining citizens' input into their decision making, but these could be improved as described in Chapters 3 and 4 and Recommendation 3.

Recommendation 2

The committee recommends that regulatory agencies adopt a risk-informed LAW system in incremental steps, relying mainly on their existing authorities under current statutes and using a four-tiered approach: (1) changes to specific facility licenses or permits and individual licensee decisions; (2) regulatory guidance to advise on specific practices; (3) regulation changes; or if necessary, (4) legislative changes.

The committee advocates a stepwise, "simplest-is-best," approach to implementing risk-informed LAW regulation and management. Acting under their existing authorities, regulatory agencies and site operators can effect significant changes from the bottom up, beginning with changes to specific facility licenses, permits, or decisions. The balance among these approaches is best determined by the agencies with the authority for regulating LAW.

By changing licenses and permits, the burden of moving toward risk-informed practices is shared by generators, facility operators, and regulators. This includes characterizing waste and providing information to the public in advance of regulatory requirements. Good business practices can lead generators toward better waste prevention, minimization, and segregation if there is more flexibility in selecting options for managing and disposing wastes.

Effective changes can be made with regulatory guidance, regulations, and new legislation. Regulatory guidance is often developed to provide specific advice regarding practices or interpretation of regulations that define acceptable conditions or requirements. Examples include Branch Technical Positions and Regulatory Guides promulgated by the USNRC.

Regulations are promulgated to implement controlling laws and statutes. Changes are often small but may occasionally result from larger initiatives. In addition, agencies can and do enter into Memoranda of Understanding (MOUs) to better align and clarify requirements where there is a shared regulatory responsibility. One example is the MOU between the USNRC and EPA on decommissioning requirements for sites containing both radioactive and hazardous materials.[5]

At the highest level of the four-tiered approach, new legislation should be targeted carefully to address a range of issues and should be balanced against the need for consistency and minimal disruption to established practices in the industry. For example, the Energy Policy Act of 2005 amended the AEA's definition of byproduct material, which will lead to more consistent regulation of materials that pose similar risks.

Recommendation 3

The committee recommends that government agencies continue to explore ways to improve their efforts to gather knowledge and opinions from stakeholders, particularly the affected and interested publics, when making LAW risk management decisions. Public stakeholders play a central role in a risk-informed decision process.

When those affected by a decision are involved in the decision-making process, the outcome is generally more accepted and more easily implemented than it would be otherwise. Management and disposal of LAW and other potential environmental hazards have evolved beyond ex post facto announcements by facility operators and regulatory agencies into a deliberative process involving partnerships with the affected and interested publics.

The committee is aware of and endorses the public outreach efforts of the DOE, EPA, USNRC, and other governmental and private organizations.[6, 7] As noted in Chapter 3 and as recognized by the committee, efforts to better include public input into risk conversations have been increasing

---

[5]The MOU between EPA and USNRC on site decommissioning is available at http://www.nrc.gov/reading-rm/doc-collections/news/2002/mou2fin.pdf.

[6]For example, the National Mining Association, a trade organization that held its Uranium Workshop in Denver, Colorado, May 24-25, 2005, devoted much of the workshop to public comments on its site remediation work. The workshop included a presentation on the status of this report.

[7]The USNRC Division of Waste Management and Environmental Protection Decommissioning Workshop held on April 20 and 21, 2005, as part of the USNRC staff's initiative to continually improve the licensing process for decommissioning sites and terminating USNRC licensees in accordance with 10 CFR 20, Subpart E.

since the 1980s. Nevertheless, during the course of this study, the committee was aware of persistent public concern with essentially all aspects of radioactive waste management.

There are a number of fundamental problems with current LAW regulatory and management practices that impede communication among affected and interested publics and those responsible for LAW that can be alleviated by implementing a risk-informed system. As noted previously, the current origin-based system is rigid and hard to understand. With risk- rather than origin-based regulatory classification as the primary consideration, a risk-informed system can give a much clearer signal that experts are making a sincere effort to ensure safety and consistency in their practices. Expert judgment is not discounted in a risk-informed system, but other diverse knowledge and opinions are transparently included in decision making. Public input into deciding when a risk is as low as reasonably achievable (ALARA), as discussed in Chapter 3, is one such example.

As noted in Chapter 4, countries such as Sweden and France have been generally more successful than the United States in gaining public stakeholder support for siting LAW disposal facilities. Reasons that those stakeholders have been more supportive include greater transparency of decision making, public enfranchisement and participation in decision making, better involvement of elected local officials, and ultimately the ability of local communities to veto an initial site selection. Besides outreach, another way a few government organizations in Europe and the United States have helped public stakeholders become more central in risk decision-making processes is by helping them hire their own technical experts.

While agencies with responsibility for LAW in the United States have improved their efforts to involve stakeholders and the public in waste disposal decisions, many citizens continue to perceive those efforts as falling short of their intended goals. A continuing, and innovative effort is needed to understand and address those shortcomings. There needs to be more effort to augment activities that inform the public, such as public meetings, with those that give public and stakeholder participants a more influential and active role in the decision-making process, such as advisory committees, citizen juries, policy dialogues, and facilitated mediations. Strong efforts will be needed when implementing a risk-informed approach to ensure that stakeholders play a central role in the decision-making process.

Recommendation 4

The committee recommends that federal and state agencies continue to harmonize their regulations for managing and disposing of AEA

and non-AEA wastes so that those wastes will be controlled consistently according to their radiological hazards rather than their origins.

In the interim report's overview of LAW, the committee developed five categories that it considered inclusive of the spectrum of LAW and that helped to point out gaps and inconsistencies in present regulation and management practices. Two major inconsistencies stood out: (1) uranium-bearing wastes are subject to different controls by federal or state authorities depending on the enterprise that generated them and, in some cases, when they were generated, even though their risks are comparable; and (2) wastes defined by statute as "low-level wastes" vary widely in their radiological properties, and hence their risks.

As discussed in Chapter 2, current initiatives by Congress, regulatory authorities, and other organizations are important initial steps in rectifying them. These initiatives should continue under current regulatory authorities as described in Chapters 2 and 4 and Recommendation 2.

Recommendation 5

The committee recommends continued collaboration among U.S. and international institutions that are responsible for controlling LAW. Greater consideration of international consensus standards as bases for U.S. regulations and practices is encouraged.

Authorities in the United States can benefit from greater consideration of standards and practices developed internationally. The international community, especially the European Commission (EC) and the International Atomic Energy Agency (IAEA), is making significant progress in developing consistent, risk-based standards for managing LAW. France and Spain have built and operate special facilities for disposing of very low level wastes.

International approaches include a number of important elements of a risk-informed system. The IAEA waste classification system is based on the radiological properties of the waste rather than its origins. For very low radioactive material concentrations, EC regulations and IAEA standards provide guidelines for wastes that pose insignificant risks to be cleared or exempted from control as radioactive material. At the high end, wastes with properties similar to wastes from nuclear fuel reprocessing are classified as "high-level wastes." In the U.S. system, only wastes from reprocessing meet the legal definition of high-level waste, leaving other wastes that might pose similar risks to be defined as "greater-than-Class C low-level wastes." Uranium-bearing wastes, however, are not included in the IAEA system, which is recognized as a shortcoming. The IAEA is continuing to revise and expand its system.

If waste management technical experts and regulators develop broad agreement, members of different publics might be more trusting of their ability to ensure safe management and disposal practices. Moving toward risk-informed practices in the United States would be consistent with many international initiatives and could have the net effect of increasing stakeholder support in all countries.

References

Ackland, L. 2002. *Making a Real Killing: Rocky Flats and the Nuclear West.* Albuquerque, N.M.: University of New Mexico Press.

Averous, J. 2003. The National Radwaste Management Plan and Low-Level Radioactive Waste Management in France. Presentation by J. Averous, French Nuclear Safety Authority, to the National Research Council's Committee on Improving the Regulation and Management of Low-Activity Radioactive Wastes. Paris, France. September 22.

Barnes, M. 2003. Testimony on Low-Level Radioactive Waste Bills, SB 824 (Bivins) and HB 1567 (West) to Senate Natural Resources and House Environmental Regulation, March 25; available at http://texas.sierraclub.org/rad_waste/testimonylwv.html.

BBC Monitoring and International Reports. 2005. South Korean Government Selects First Nuclear Waste Deposit Site. Yonhap News Agency, Seoul. November 3.

Becker, B.D. et al. 2005. A Historical Look at Nevada Test Site Low-Level Waste Disposal Operations. Presentation by B.D. Becker, Bechtel Nevada, et al. at Waste Management '05. Tucson, Arizona. February 27-March 3.

Beierle, T. C., and J. Cayford. 2002. *Democracy in Practice: Public Participation in Environmental Decisions.* Washington, D.C.: Resources for the Future.

Brown, P., and E. J. Mikkelsen. 1990. *No Safe Place.* Berkeley: University of California Press.

BRWM (Board on Radioactive Waste Management). 2002. Project Prospectus: Improving the Regulation and Management of Low-Activity Radioactive Wastes.

Carlsson, T. 2002. Presentation to VALDOC Summer School, Borgholm, Sweden, June 13.

Chen, S. Y., and D. W. Moeller. 2003. Releasing "clean" or "contaminated" materials? *Health Physics News* 31: 12-13.

Dawson, S. E., P. H. Charley, and P. Harrison Jr. 1997. Advocacy and social actions among Navajo uranium workers and their families. Pp. 391-407 in *Social Work in Health Settings: Practice in Content*, 2nd ed., eds. T. Schwaber Kerson and Associates. New York: Haworth Press.

D&D Monitor (Decommissioning and Decontamination Monitor). 2005. May 2.

97

Diaz, N. J. 2005a. Draft Bill: Nuclear Safety and Security Act of 2005. Letter from Nils J. Diaz, Chairman, United States Nuclear Regulatory Commission, to the Honorable Richard B. Cheney, President of the United States Senate. March 30.

Diaz, N. J. 2005b. Proposed Rule: Radiological Criteria for Controlling the Disposition of Solid Materials. Commission Voting Record: Comments of Chairman Diaz on SECY-05-0054. United States Nuclear Regulatory Commission. June 1.

DOE (Department of Energy). 2005. Ionizing Radiation Dose Ranges. Chart compiled by the DOE Office of Biological and Environmental Research. Washington, D.C.: U.S. Department of Energy. August.

DOE. 2006 (draft). Implementation Guide for the Control and Release of Property with Residual Radioactive Material. DOE G 441.1-XX. Washington, D.C.: U.S. Department of Energy.

EC (European Commission). 1996a. Council Directive 96/29/EURATOM of 13 May 1996 laying down basic safety standards for the protection of the health of workers and the general public against the dangers arising from ionizing radiation. Luxembourg: European Commission.

EC. 1996b. *Criteria for Establishing Harmonized Categories of Waste Based on the Storage and Disposal Routes.* Report EUR 17240 EN. Luxembourg: European Commission.

EC. 2000a. *Radiation Protection 122: Practical Use of the Concepts of Clearance and Exemption.* Report RP122. Luxembourg: European Commission.

EC. 2000b. *Management and Disposal of Disused Radioactive Sources in the European Union.* Report EUR 18186 EN. Luxembourg: European Commission.

EPA (Environmental Protection Agency). 1992. Total Cost Assessment: Accelerating Industrial Pollution Prevention Through Innovative Project Financial Analysis. EPA/741/R-2/002. Washington, D.C.: EPA Office of Pollution Prevention and Toxics.

EPA. 1994. A Review of Computer Process Simulation in Industrial Pollution Prevention. EPA/600/R-94/128. Washington, D.C.: EPA Office of Research and Development.

EPA. 1995. Design for the Environment. Building Partnerships for Environmental Improvement. EPA/600/K-93/002. Washington, D.C.: EPA Office of Pollution Prevention and Toxics.

EPA. 1997. Developing and Using Production-Adjusted Measurements of Pollution Prevention. EPA/600/R-97/048. Washington, D.C.: EPA Office of Research and Development.

EPA. 2003. Approaches to an Integrated Framework for Management and Disposal of Low-Activity Radioactive Waste: Request for Comment; Proposed Rule. 40 CFR Chapter 1. *Federal Register*, Vol. 68, No. 222, pp. 65119-65151. Tuesday, November 18.

EPA. 2005a. Public Information and Input on WIPP. 2005 EPA WIPP Recertification Fact Sheet No. 2, available at http://www.epa.gov/radiation/docs/wipp/recertification/fs2-getinvolved.pdf.

EPA. 2005b. 2005 EPA WIPP Recertification Fact Sheet No. 6, available at http://www.epa.gov/radiation/docs/wipp/recertification/fs6-karst.pdf.

Federline, M. 2004. Management and Disposal Strategies for Low-Activity Waste in the United States. Presentation by Margaret V. Federline Deputy Director, Office of Nuclear Materials Safety and Safeguards, U.S. Nuclear Regulatory Commission at the International Atomic Energy Agency Meeting on Low-Activity Wastes. Cordoba, Spain. December 13-17.

Friedman, S. M. 1981. Blueprint for breakdown: Three Mile Island and the media before the accident. *Journal of Communication* spring: 116-128.

Friedman, S. M. 1991. Radiation risk reporting: How well is the press doing its job? *USA Today* 120: 78-81.

Friedman, S.M. 2001. Risk communication, the Hanford Thyroid Disease Study and draft reports. *Risk: Health Safety & Environment* spring: 91-105.

GAO (Government Accountability Office). 1999. Low-Level Radioactive Wastes: States Are Not Developing Disposal Facilities. GAO/RCED-99-238. Washington, D.C.: Government Accountability Office.

GAO. 2004. Low-Level Radioactive Waste: Disposal Availability Adequate in the Short Term, but Oversight Needed to Identify Any Future Shortfalls. GAO-04-604. Washington, D.C.: Government Accountability Office.

GAO. 2005. Department of Energy: Improved Guidance, Oversight, and Planning Are Needed to Better Identify Cost-Saving Alternatives for Managing Low-Level Radioactive Waste. GAO-06-94. Washington, D.C.: Government Accountability Office.

Garrick, B. J., and S. Kaplan. 1995. Radioactive and Mixed Waste—Risk as a Basis for Waste Classification. NCRP Symposium, Proceedings No. 2, pp. 59-73. Bethesda, Md.: National Council on Radiation Protection and Measurements.

Genoa, P. 2003. Milestones and Millstones: Industry Experience with Low-Activity Waste Disposals. Presentation by Paul Genoa, Nuclear Energy Institute, to the National Research Council's Committee on Improving the Regulation and Management of Low-Activity Radioactive Wastes. Washington, D.C. June 12.

Goldammer, W. 2004. Regulatory approach to TENORM in Germany. Pp. 133-137 in *Proceedings of the Cordoba Conference on the Disposal of Low Activity Radioactive Waste*, IAEA-CN-124. Vienna: International Atomic Energy Agency.

Hirusawa, S. 2004. Values of Activity Concentration for the Clearance of Bulk Amounts of Materials. Presentation by Shingenobu Hirusawa, Institute of Applied Energy (Japan), to the National Research Council's Committee on Improving the Regulation and Management of Low-Activity Radioactive Wastes. Washington, D.C. November 30.

IAEA (International Atomic Energy Agency). 1987. Techniques and Practices for Pre-Treatment of Low- and Intermediate-Level Radioactive Solid and Liquid Wastes, Technical Reports Series No. 272. Vienna: IAEA.

IAEA. 1988. The Radiological Accident in Goiania, available at http://www-pub.iaea.org/MTCD/publications/PDF/Pub815_web.pdf.

IAEA. 1994. Classification of Radioactive Waste. Safety Series No. 111-G-1.1. Vienna: IAEA.

IAEA. 1996. International Basic Safety Standards for Protection Against Ionizing Radiation and for the Safety of Radiation Sources. Safety Series No. 115. Vienna: IAEA.

IAEA. 1997. Joint Convention of the Safety of Spent Fuel Management and on the Safety of Radioactive Waste Management. INFCIRC/516. Vienna: IAEA.

IAEA. 1998. IAEA Conference on Safety of Radiation Sources and Security of Radioactive Materials, Dijon, France, September 14-18.

IAEA. 1999. Near Surface Disposal of Radioactive Waste, Safety Requirement WS-R-1. Vienna: IAEA.

IAEA. 2000. International Conference of National Regulatory Authorities with Competence in the Safety of Radiation Sources and the Security of Radioactive Materials, Buenos Aires, Argentina, December 11-15.

IAEA. 2001. International Conference on Security of Material: Measures to Detect, Intercept and Respond to the Illicit Uses of Nuclear Material and Radioactive Sources, Stockholm, Sweden, May 7-11.

IAEA. 2002. Management of Radioactive Waste from the Mining and Milling of Ores. Safety Guide. Safety Standards Series No. WS-G-1.2. Vienna: IAEA.

IAEA. 2003a. International Conference on the Security of Radioactive Sources. Vienna: IAEA, March 10-13.

IAEA. 2003b. Considerations in the Development of Near Surface Repositories for Radioactive Waste. Technical Reports Series No. 417. Vienna: IAEA.

IAEA. 2003c. Scientific and Technical Basis for the Geological Disposal of Radioactive Wastes. Technical Reports Series No. 413. Vienna: IAEA.

IAEA. 2004a. The IAEA Code of Conduct on the Safety and Security of Radioactive Sources, IAEA/CODEOC/2004. Vienna: IAEA.

IAEA. 2004b. Application of the Concepts of Exclusion, Exemption and Clearance. Safety Guide No. RS-G-1.7. Vienna: IAEA.

IAEA. 2005a. IAEA Guidance on the Import and Export of Radioactive Sources, IAEA/CODEOC/IMP-EXP/2005. Vienna: IAEA.

IAEA. 2005b. International Conference on the Safety and Security of Radioactive Sources: Towards a Global System for the Continuous Control of Sources Throughout their Life Cycle, Bordeaux, France, June 27-July 1.

ICRP (International Commission on Radiological Protection). 1998. Radiation protection recommendations as applied to the disposal of long-lived solid radioactive waste. *Annals of the ICRP* 28(4): i-vii, 1-25.

INAC (Indian and Northern Affairs Canada). 2003. Canadian Arctic Contaminants Assessment Report II. Northern Contaminants Program. Ottawa, Ontario: INAC.

Jablonski, S. 2004. The Texas LLRW Disposal License Review Process. Presentation by Susan Jablonski, Texas Commission on Environmental Quality, at the 19th Annual International Radioactive Exchange LLRW Decisionmakers' Forum and Technical Symposium. Midland, Texas. November 17.

Johnsrud, J. 2005. Low-Activity Wastes in Our Backyards: Views of Stakeholders. Presentation by Judith Johnsrud, Chair, National Nuclear Waste Committee, Sierra Club at the American Association for the Advancement of Science Symposium: Low-Activity Radioactive Waste—The Problem Is Everywhere. Washington, D.C. February 21.

Kaplan, S., and B. J. Garrick. 1981. On the quantitative definition of risk. *Risk Analysis* 1(1): 11-27.

Krimsky, S., and A. Plough. 1988. *Environmental Hazards: Communicating Risks as a Social Process*. Dover, Mass.: Auburn House.

Leroy, D. 2004. Improving the Regulation and Management of Low-Activity Radioactive Wastes in the United States. Presentation at the International Atomic Energy Agency Symposium on the Disposal of Low Activity Radioactive Waste. Cordoba, Spain. December.

LLW Forum (Low-Level Radioactive Waste Forum, Inc.). 2005. Discussion of Issues: Management of Commercial Low-Level Radioactive Waste, available at http://www.llwforum.org/pdfs/Forumpolicy9-22-05FINALFORPUBLICATION_3_.pdf.

Lucas, G. 2002. Nuclear waste rule angers critics: Environmentalists fight state law allowing disposal of small amounts in city dumps. *San Francisco Chronicle*, March 19, p. A-13.

MacGregor, D. G., J. Flynn, P. Slovic, and C. K. Mertz. 2002. Perception of Radiation Exposure. Part 1. Perception of Risk and Judgment of Harm. Report No. 02-03. Eugene, Ore.: Decision Research.

Markstrom, C. A., and P. H. Charley. 2001. Psychological effects of technological/human-caused environmental disasters: Examination of Navajos and uranium. American Psychology, the American Indian & Alaskan Native Mental Health Research: A Journal of the National Center.

McDaniel, T. 2004. U.S. Army Corps of Engineers: Perspective on Low-Activity Waste. Presentation by Tomiann McDaniel, U.S. Army Corps of Engineers, to the National Research Council's Committee on Improving the Regulation and Management of Low-Activity Radioactive Wastes. Washington, D.C. November 30.

Merrifield, J. S. 2005. Proposed Rule: Radiological Criteria for Controlling the Disposition of Solid Materials. Commission Voting Record: Comments of Chairman Commissioner Merrifield on SECY-05-0054. United States Nuclear Regulatory Commission. May 5.

Meserve, R. 2005. Low-Activity Radioactive Waste: An Increasingly Complex Dilemma for Science and Society. Presentation at the American Association for the Advancement of Science Symposium: Low-Activity Radioactive Waste—The Problem is Everywhere. Washington, D.C. February 21.

Murray, F. E. S., and R. H. K. Vietor. 1995. Xerox: Design for the Environment. Harvard Business School case 9-794-022, May.

NCRP (National Council on Radiation Protection and Measurements). 1987. Ionizing Radiation Exposure to the Population of the United States. NCRP Publication 93. Bethesda, Md.: National Council on Radiation Protection and Measurements.

NCRP. 2002. Risk-Based Classification of Radioactive and Hazardous Chemical Wastes. NCRP Publication 139. Bethesda, Md.: National Council on Radiation Protection and Measurements.

NRC (National Research Council). 1983. *Risk Assessment in the Federal Government: Managing the Process.* Washington, D.C.: National Academy Press.

NRC. 1984. *Social and Economic Aspects of Radioactive Waste Disposal: Considerations for Institutional Management.* Washington, D.C.: National Academy Press.

NRC. 1994. *Science and Judgment in Risk Assessment.* Washington, D.C.: National Academy Press.

NRC. 1996. *Understanding Risk.* Washington, D.C.: National Academy Press.

NRC. 1999. *Evaluation of Guidelines for Exposures to Technologically Enhanced Naturally Occurring Radioactive Materials.* Washington, D.C.: National Academy Press.

NRC. 2001a. *The Impact of Low-Level Radioactive Waste Management Policy on Biomedical Research in the United States.* Washington, D.C.: National Academy Press.

NRC. 2001b. *A Risk Management Strategy for PCB-Contaminated Sediments.* Washington, D.C.: National Academy Press.

NRC. 2001c. *Disposition of High-Level Waste and Spent Nuclear Fuel: The Continuing Societal and Technical Challenges.* Washington, D.C.: National Academy Press.

NRC. 2002. *The Disposition Dilemma: Controlling the Release of Solid Materials from Nuclear Regulatory Commission-Licensed Facilities.* Washington, D.C.: National Academy Press.

NRC. 2003a. *Improving the Regulation and Management of Low-Activity Radioactive Wastes. Interim Report on Current Regulations, Inventories, and Practices.* Washington, D.C.: The National Academies Press.

NRC. 2003b. *One Step at a Time: The Staged Development of Geologic Repositories for High-Level Radioactive Waste.* Washington, D.C.: The National Academies Press.

NRC. 2005a. *Health Risks from Exposure to Low Levels of Ionizing Radiation.* Washington, D.C.: The National Academies Press.

NRC. 2005b. *Risk and Decisions About the Disposition of Transuranic and High-Level Radioactive Waste.* Washington, D.C.: The National Academies Press.

NRC. 2005c. *Tank Wastes Planned for Onsite Disposal at Three Department of Energy Sites: The Savannah River Site.* Interim Report. Washington, D.C.: The National Academies Press.

NSC (Nuclear Safety Commission of Japan). 2004. Commonly Important Issues for the Safety Regulations of Radioactive Waste Disposal (in English). June 2004.

Oskarshamn. 2003. Site Investigation in the Municipality of Oskarshamn: Decision on Site Investigation. Decision Statement, available at http://web.wpab.se/lko/Data/public-se/en_dokument.asp?MainTyp=1. Last accessed February 2, 2006.

OTA (Office of Technology Assessment). 1992. Green Products by Design: Choices for a Cleaner Environment. OTA-E-541. Washington, D.C.: U.S. Government Printing Office.

Pasternak, A. 2003. Perspective: A National Solution for a National Problem. Radwaste Solutions. American Nuclear Society. La Grange Park, Illinois. September/October.

Phoenix Environmental. 2001. Evaluation of the U.S. Environmental Protection Agency's Public Outreach Program During the Certification Process at the Waste Isolation Pilot Plant in New Mexico, available at http://www.epa.gov/radiation/docs/wipp/wipp_cert_eval_0401_execsum.pdf.

Public Citizen. 2005. Stop the NRC from Authorizing the Release of Radioactive Waste into Communities! March 31, 2005, available at http://www.citizen.org/documents/ACF65EC.pdf.

Risk Commission (U.S. Commission on Risk Assessment and Risk Management). 1997. Final Report, Volume 1. Framework for Environmental Health Risk Management. GPO #055-000-00567-2, available at http://www.riskworld.com/Nreports/1996/risk_rpt/Rr6me001.htm.

Rondinelli, D. A., and M. A. Berry. 2000. Corporate environmental management and public policy: Bridging the gap. *American Behavioral Scientist* 44(2): 168-187.

Schneider, K. 1988. How secrecy on atomic weapons helped breed a policy of disregard. *New York Times*, Nov. 13, p. E7.

Slovic, P. 2000. Perception of risk from radiation. Pp. 264-274 in *The Perception of Risk*, ed. P. Slovic. London: Earthscan Publications.

USNRC (U.S. Nuclear Regulatory Commission). 1998. White paper on risk-informed and performance-based regulation. SECY-98-144. Memorandum to the Commissioners from L. Joseph Callan, Executive Director for Operations. Washington, D.C.: USNRC.

USNRC. 2002. NRC Public Meetings (brochure) NUREG/BR-0297, available at http://www.nrc.gov/reading-rm/doc-collections/nuregs/brochures/br0297/br0297.pdf.

USNRC. 2005. Proposed Rule: Radiological Criteria for Controlling the Disposition of Solid Materials. SECY-05-0054. Memorandum to Luis A. Reyes from Annette L. Vietti-Cook. Washington, D.C.: USNRC.

West, G. E. 2004. A Briefing on the Proposed WCS Disposal Site. Presentation by the Honorable G. E. (Buddy) West, Member Texas House of Representatives, 81st District, at the 19th Annual International Radioactive Exchange LLRW Decisionmakers' Forum and Technical Symposium. Midland, Texas. November 17.

Wiley, J. 2005. Why we need better management of low-activity radioactive wastes. *Radwaste Solutions* 21(3) May/June.

Yankelovich, D. 1991. *Coming to Public Judgment. Making Democracy Work in a Complex World.* Syracuse, N.Y.: Syracuse University Press.

Zuloaga, P. 2003. Low-Level Radioactive Waste Management in Spain. Presentation by P. Zuloago, Empresa Nacional de Residuos Radioactivos, SA, to the National Research Council's Committee on Improving the Regulation and Management of Low-Activity Radioactive Wastes. Paris, France. September 22.

Appendixes

Appendix A: Interim Report

Executive Summary

L ow-activity radioactive wastes include a broad spectrum of materials for which a regulatory patchwork has evolved over almost 60 years. These wastes present less of a radiation hazard than either spent nuclear fuel or high-level radioactive waste.[1] Low-activity wastes, however, may produce potential radiation exposure at well above background levels and if not properly controlled may represent a significant chronic (and, in some cases, an acute) hazard.[2] For some low-activity wastes the present system of controls may be overly restrictive, but it may result in the neglect of others that pose an equal or higher risk.

The purpose of this interim report is to provide an overview of current low-activity waste regulations and management practices (see Sidebar ES.1). In developing this overview, the committee[3] has sought to identify gaps and inconsistencies that suggest areas for improvements. This initial fact-finding phase of the project led the committee to the findings that conclude this interim report. The committee will assess options for improving the current practices and provide recommendations in its final report.

In initiating this study, the Board on Radioactive Waste Management used the term "low-activity waste" to denote a spectrum of radioactive

---

[1] See Disposition of High-Level Waste and Spent Nuclear Fuel: The Continuing Societal and Technical Challenges (NRC, 2001a) and One Step at a Time: The Staged Development of Geologic Repositories for High-Level Radioactive Waste (NRC, 2003).

[2] See Health Effects of Exposure to Low Levels of Ionizing Radiation: BEIR V (NRC, 1990).

[3] The Committee on Improving Practices for Regulating and Managing Low-Activity Radioactive Waste is referred to as "the committee" throughout this report.

SIDEBAR ES.1
Purpose of This Report

This study was initiated by the National Academies' Board on Radioactive Waste Management. Due to financial constraints, the study was divided into two phases. This interim report, which concludes phase one, addresses current low-activity waste regulations and practices according to the following parts of the study's task statement:

(1) Using available information from public domain sources, provide a summary of the sources, forms, quantities, hazards, and other identifying characteristics of low-activity waste in the United States; and
(2) review and summarize current policies and practices for regulating, treating, and disposing of low-activity waste, including the quantitative (including risk) bases for existing regulatory systems, and identify waste streams that are not being regulated or managed in a safe or cost effective manner.

Phase two will assess options for improving regulations and practices (see Chapter 1, Sidebar 1.1) and provide a final report.

materials declared as wastes from a variety of activities—national defense, nuclear power, industry, medicine, research, and mineral recovery.[4] Given this broad charter, the committee sought to develop a concise list of categories that would include low-activity wastes from essentially all sources,[5] yet by focusing on their inherent radiological properties rather than their origins, emphasize gaps and inconsistencies between their current regulation and management and their actual radiological hazards. The committee agreed that the following is an instructive and inclusive categorization of the wastes to be addressed:

• Wastes containing types and quantities of radioactive materials that fall well within the Nuclear Regulatory Commission (USNRC) classi-

---

[4]The Board intended the term "low-activity waste" to be more inclusive than "low-level waste," which has a specific definition under the Nuclear Waste Policy Act (see Chapter 2). The term "low-activity waste" has sometimes been applied to the lower activity fractions of Department of Energy (DOE) tank waste. The committee does not use the term in this sense.
[5]The committee did not include waste containing only short-lived radioactivity (on the order of a year or less), which simply decays away during storage. These wastes do not present long-term management or disposal challenges.

fication system for low-level waste, e.g., Class A, B, and C (see Chapters 2 and 3 and Appendix A). These include wastes from nuclear utilities, other industries, medicine, and research, which are disposed in USNRC-licensed, commercially operated facilities ("commercial low-level waste"), and similar wastes produced and disposed at DOE sites ("defense low-level waste").

• Slightly radioactive solid materials—debris, rubble, and contaminated soils from nuclear facility decommissioning and site cleanup. They arise in very large volumes but produce very low or practically undetectable levels of radiation. They fall at the very bottom of USNRC Class A (the lowest of the classes).

• Discrete sources—out-of-service radiation sources and associated materials from industrial, medical, and research applications. Although defined by statute as low-level waste, they may emit high enough levels of radiation to cause acute effects in humans or serious contamination incidents. Larger sources may exceed USNRC Class C (the highest of the classes).

• Uranium and thorium ore processing wastes. These wastes have been produced in large volumes from the recovery of uranium and thorium for nuclear applications. Their radiological hazards arise not only from radioactive uranium and thorium isotopes, but also from their radioactive decay products, especially radium, which can migrate into drinking water, and radon, which is a gas.

• Naturally occurring and technologically enhanced naturally occurring radioactive materials (NORM and TENORM) wastes. These wastes arise coincidentally from the recovery of natural resources (extraction of rare earth minerals and other mining operations, oil, and gas) and water treatment. Like uranium and thorium wastes, they arise in large volumes and their radiological hazards result from uranium, thorium, and their radioactive decay products, radium and radon.

Throughout this report the committee will use these categories to illustrate gaps and inconsistencies in the current regulations for wastes with very different levels of radioactivity, volumes, and radioactive half-lives; and inconsistencies in regulating wastes that are radiologically similar to each other.

At least 12 federal statutes apply to low-activity wastes. The broadest of these is the Atomic Energy Act (AEA), which defines wastes in the first four categories listed above as "byproduct" materials and provides federal authority for their regulation. Wastes in the first three categories meet the definition of low-level waste (LLW) given in the Nuclear Waste Policy Act (NWPA) of 1982, as amended. The NWPA provides no statutory upper or lower limit on the radioactivity in LLW. Uranium- and thorium-contaminated wastes produced after the Uranium Mill Tailings Radiation Control Act (UMTRCA) was passed in 1978 must be disposed in licensed radioactive waste facilities.[6] There are more disposal options for uranium- and thorium-contaminated wastes produced prior to UMTRCA, which are managed under the Formerly Utilized Sites Remedial Action Program (FUSRAP). Thus the disposal options for FUSRAP and UMTRCA wastes differ even though the materials are the same (or similar).

LLW generated or disposed in the commercial sector are regulated by the USNRC under its authority to license nuclear facilities and the possession of nuclear materials. The Environmental Protection Agency (EPA) has authority to regulate environmental radiation exposure as well as hazardous chemical wastes. Wastes that contain both radionuclides and hazardous chemicals are referred to as "mixed wastes" and may be subject to regulation by both the USNRC and EPA. The DOE is self-regulating for defense wastes on its own sites. The Department of Transportation regulates the shipment of radioactive materials while the USNRC has the authority to regulate certain packages for transportation of nuclear materials.

The states have three important responsibilities with regard to low-activity wastes:

1. The Low-Level Radioactive Waste Policy Act of 1980, as amended, makes each state responsible for disposing of its own LLW and encourages the formation of state compacts (congressionally ratified agreements among groups of states) for providing disposal facilities.[7]

2. States may assume portions of the USNRC's regulatory authority by becoming a USNRC Agreement State. Thirty-three states are Agreement States, including the three that currently host LLW disposal facilities (South Carolina, Utah, and Washington).

3. The states regulate non-AEA wastes because these wastes are not covered by federal statutes. An especially important role for the states is the regulation of NORM and TENORM wastes from a number of activities, including mining, oil and gas production, and water treatment.

Of the wastes described in this interim report, LLW from DOE and commercial nuclear facilities have received the most attention from regulators and the public. LLW in the form of debris, rubble, and contaminated soils from facility decommissioning and site cleanup constitutes

---

[6]Strictly speaking, UMTRCA also applies to wastes at facilities licensed by the USNRC before 1978 (see Chapter 2).

[7]As discussed in Chapter 3, the Act did not lead to establishment of new disposal sites as intended.

much larger volumes than LLW from operational facilities, but it generally contains very little radioactive material. Conversely, discrete radioactive sources that are no longer useful also meet the definition of LLW although they may contain highly concentrated radioactive materials.

Millions of cubic meters of tailings and other wastes from mining and processing uranium and thorium ores are stored or disposed in piles near their origin. Like LLW, uranium and thorium wastes are subject to the AEA, but concern about them comes mainly from citizens living near these wastes. NORM and TENORM wastes contain the same long-lived radioactive constituents as uranium and thorium wastes and arise in equally large or larger volumes. NORM and TENORM wastes are not subject to the AEA, and there is less consistency in their regulation and little public concern about them.

FINDINGS

In general, the committee believes that there is adequate statutory and institutional authority to ensure safe management of low-activity wastes, but the current patchwork of regulations is complex and inconsistent— which has led to instances of inefficient management practices and possibly in some cases increased risk overall. Existing authorities have not been exercised consistently for some wastes. The system is likely to grow less efficient if the patchwork approach to regulation continues in the future.

Finding 1

Current statutes and regulations for low-activity radioactive wastes provide adequate authority for protection of workers and the public.

In its fact-finding meetings, site visits, and review of relevant literature, the committee found no instances where the legal and regulatory authority of federal and state agencies was inadequate to protect human health. This finding is consistent with previous studies by the National Academies and the National Council on Radiation Protection and Measurements (NCRP) (NRC, 1999a, 2002a; NCRP, 2002). Some states, however, have chosen not to exercise regulatory authority over NORM and TENORM wastes. The USNRC has determined not to regulate certain pre-1978 uranium and thorium wastes. The EPA has so far not exercised its authority under the Toxic Substance Control Act to regulate non-AEA radioactive wastes. In addition, some wastes have not been adequately controlled in spite of the existence of regulatory authority. Incidents in which out-of-use sealed sources were melted with scrap steel have been expensive, led to very conservative practices in the steel and nuclear industries, and fueled public distrust in the regulatory system (NRC, 2002a; HPS, 2002; Turner, 2003).

Finding 2

The current system of managing and regulating low-activity waste is complex. It was developed under a patchwork system that has evolved based on the origins of the waste.

In its information-gathering the committee received a clear message from agencies responsible for managing and regulating low-activity waste: A more consistent, simpler, performance-based and risk-informed approach to regulation is needed (see Chapter 4, Sidebar 4.3). Many committee members themselves had difficulty in understanding the regulations well enough to discuss the system and its applications. Similarly the NCRP found that the current waste classification systems "are not transparent or defensible" and that the "classification systems are becoming increasingly complex as additional waste streams are incorporated into the system" (NCRP, 2002, p. 65).

Findings 3 and 4

Certain categories of low-activity wastes have not received consistent regulatory oversight and management.

Current regulations for low-activity wastes are not based on a systematic consideration of risks.

Regulations focused on the wastes' origins have led to inconsistencies relative to their likely radiological risks. NORM and TENORM generally are not regulated by federal agencies, and state regulation of these wastes is not consistent. Nevertheless, these wastes may have significant concentrations of radioactive materials as compared to some highly regulated waste streams (e.g., from the nuclear industry). As described in Chapter 4, NORM wastes routinely accepted at a landfill triggered a radiation monitor intended to ensure that rubble from a decommissioned nuclear reactor meets very strict limits on its radioactivity.

Uranium mining and processing wastes, which are radiologically similar to NORM wastes, are regulated by their date of origin. Federal regulations do not prohibit ore processing residuals at facilities that were not under license by the USNRC before the 1978 passage of UMTRCA from being disposed in landfills. However, mill tailings generated after UMTRCA must be disposed in licensed radioactive waste facilities.

In addition to inconsistencies in regulating the radiological risks, current regulations generally overlook trade-offs between radiological and nonradiological risks. Very large (100,000 cubic meter) volumes of slightly contaminated soil and debris, and very heavy nuclear reactor components are being transported long distances for disposal. In developing current requirements for how low-activity wastes are managed or disposed, worker risks in excavating, loading, and unloading large-volume wastes; risks of transportation accidents; and environmental risks and costs (e.g., consuming large amounts of fossil fuel) have not been analyzed and compared in a systematic way to radiological risks.

PUBLIC CONCERNS REGARDING LOW-ACTIVITY WASTES: AN ISSUE FOR THE FINAL REPORT

On beginning this study, the committee was aware that there is persistent and widespread public concern with all aspects of radioactive waste management and disposal (NRC, 1996, 2001a, 2002a, 2003; GAO, 1999; Dunlap et al., 1993). During the committee's open sessions, members of the attending public expressed considerable lack of trust in the low-activity waste regulatory system due to its complexity, inflexibility, and inconsistency. These factors have apparently raised doubts about the current system's capability for protecting public health.

The task of this interim report was to develop an overview of current regulatory and management practices for low-activity waste, and thus set the stage for the committee's final report, which will assess policy and technical options for improving the current practices. The assessments will include risk-informed options, and the committee strongly believes that issues of public trust and risk perception will be important considerations in the final report.

Appendix A: Interim Report

1
Introduction

This study was initiated by the National Academies' Board on Radioactive Waste Management (the Board), which observed that statutes and regulations administered by the state and federal agencies that control low-activity wastes have developed in an ad hoc manner over almost 60 years. They usually reflect the waste's origin from national defense, nuclear power, industrial, institutional, or natural sources rather than its radiological hazard. Inconsistencies in the regulatory patchwork or its application have led to very restrictive controls for some low-activity wastes but the relative neglect of others.

The purpose of this interim report is to provide an overview of current regulations and management practices, in conformance with items 1 and 2 of the project's task statement (see Sidebar 1.1). In developing the overview, the committee[1] has sought to identify gaps and inconsistencies that would suggest areas for significant improvements. This initial fact-finding phase of the project led the committee to the findings that conclude this report. The committee will address item 3 of the task statement and provide recommendations in its final report.

WHAT ARE LOW-ACTIVITY RADIOACTIVE WASTES?

In initiating this study, the Board used the term "low-activity waste" to denote a spectrum of radioactive materials declared as wastes from a

---

[1]The Committee on Improving Practices for Regulating and Managing Low-Activity Radioactive Waste is referred to as "the committee" throughout this report.

SIDEBAR 1.1
Task Statement

The objective of this study is to evaluate options for improving practices for regulating and managing low-activity radioactive waste in the United States. The study will focus on the following three tasks:

1. Using available information from public domain sources, provide a summary of the sources, forms, quantities, hazards, and other identifying characteristics of low-activity waste in the United States;

2. Review and summarize current policies and practices for regulating, treating, and disposing of low-activity waste, including the quantitative (including risk) bases for existing regulatory systems, and identify waste streams that are not being regulated or managed in a safe or cost-effective manner; and

3. Provide an assessment of technical and policy options for improving practices for regulating and managing low-activity waste to enhance technical soundness, ensure continued protection of public and environmental health, and increase cost effectiveness. This assessment should include an examination of options for utilizing risk-informed practices for identifying, regulating, and managing low-activity waste irrespective of its classification.

---

variety of national defense and private-sector activities.[2] These low-activity wastes generally contain lower levels of radioactive material and present less of a hazard to public and environmental health than either spent nuclear fuel or high-level waste from chemical processing of spent fuel, both of which are highly hazardous and tightly regulated.[3] However, low-activity wastes may contain naturally occurring or other long-lived radionuclides at well above background levels, and it may represent a significant chronic (and, in some cases, an acute) hazard to public and environmental health.[4]

---

[2]The Board intended the term "low-activity waste" to be more inclusive than "low-level waste," which has a specific definition under the Nuclear Waste Policy Act (see Chapter 2). The term "low-activity waste" has sometimes been applied to the lower activity fractions of Department of Energy (DOE) tank waste. The committee does not use the term in this sense.

[3]See Disposition of High-Level Waste and Spent Nuclear Fuel: The Continuing Societal and Technical Challenges (NRC, 2001a) and One Step at a Time: The Staged Development of Geologic Repositories for High-Level Radioactive Waste (NRC, 2003). Transuranic wastes, which are controlled by the DOE, are addressed in several other National Research Council reports (NRC, 2001b, 2002b, 2002c) and are not included in this study.

[4]See Health Effects of Exposure to Low Levels of Ionizing Radiation: BEIR V (NRC, 1990).

Given this broad charter, the committee sought to develop a concise list of categories that would include low-activity wastes from essentially all sources,[5] yet by focusing on their inherent radiological properties rather than their origins, emphasize gaps and inconsistencies between their current regulation and management and their actual radiological properties. The committee agreed that the following is an instructive and inclusive categorization of the wastes to be addressed:

• Wastes containing types and quantities of radioactive materials that fall well within the Nuclear Regulatory Commission (USNRC) classification system for low-level waste, e.g., Class A, B, and C (see Chapters 2 and 3 and Appendix A). These include wastes from nuclear utilities, other industries, medicine, and research, which are disposed in USNRC-licensed, commercially operated facilities ("commercial low-level waste"), and similar wastes produced and disposed at Department of Energy (DOE) sites ("defense low-level waste").

• Slightly radioactive solid materials—debris, rubble, and contaminated soils from nuclear facility decommissioning and site cleanup. They arise in very large volumes but produce very low or practically undetectable levels of radiation. They fall at the very bottom of USNRC Class A (the lowest of the classes).

• Discrete sources—out-of-service radiation sources and associated materials from industrial, medical, and research applications. Although defined by statute as low-level waste, they may emit high enough levels of radiation to cause acute effects in humans or serious contamination incidents.[6] Larger sources may exceed USNRC Class C (the highest of the classes).

• Uranium and thorium ore processing wastes. These wastes have been produced in large volumes from the recovery of uranium and thorium for nuclear applications. Their radiological hazards arise not only from the radioactive uranium and thorium isotopes, but also from their radioactive decay products, especially radium, which can migrate into drinking water, and radon, which is a gas.

• Naturally occurring and technologically enhanced naturally occurring radioactive materials (NORM and TENORM) wastes. These wastes arise coincidentally from the recovery of natural resources (extrac-

---

[5]The committee did not include waste containing only short-lived radioactivity (on the order of a year or less), which simply decays away during storage. These wastes do not present long-term management or disposal challenges.

[6]For completeness, radium sources and accelerator-produced material can be included in this category although they do not meet the statutory definition of low-level waste (see Chapter 2).

tion of rare earth minerals and other mining operations, oil, and gas) and water treatment. Like uranium and thorium wastes, they arise in large volumes and their radiological hazards result from uranium, thorium, and their radioactive decay products, radium and radon.

As will be discussed later in this report, wastes in the first four categories fall under the Atomic Energy Act, which provides authority for their control by federal agencies. Wastes in the first three categories all meet the statutory definition of low-level waste, although their physical and radiological properties, and hence their hazards, vary greatly. Wastes in the last two categories are similar in their physical and radiological properties, but the federal government has regulatory authority over the former and the states have authority over the latter. Table 1.1 summarizes the committee's categorization of low-activity wastes.

APPROACH TO THE TASK STATEMENT

In developing its overview of current inventories, regulations, and management practices for this interim report (parts 1 and 2 of the task statement), the committee encountered a massive amount of literature on federal and state regulations, inventory data, and management practices. This report does not attempt to replicate the detailed information already available; rather, the report summarizes the information that led to the committee's findings and points to possible improvements in the overall regulatory structure, which the committee will examine in its final report (part 3 of the task statement).

Information Sources

The main sources of information for this interim report included:

- Information-gathering meetings and site visits,
- Previously published studies, and
- Internet material.

First-hand information was provided to the committee at five information-gathering meetings and three site visits. This information was presented by the study sponsors, representatives of other regulatory and operating organizations, local officials, and members of the public. The committee held its first information-gathering meeting in Washington, D.C. on December 4-5, 2002, to receive presentations from study sponsors and comments from other interested individuals. Information-gathering and site visits included Richland, Washington (Hanford and U.S.

TABLE 1.1 The Committee's Categorization of Low-Activity Wastes

Category	Principal Origins	Typical Examples
Defense and commercial low-level waste	Operations at DOE sites, nuclear power and research reactors, medical facilities	Trash, equipment, construction materials, process residues, soils
Slightly radioactive solid materials	Decommissioning of nuclear facilities at DOE and civilian sites, and site cleanup	Debris, rubble, construction materials, soils
Discrete sources	Applications of radiation sources in industry, medicine, and research	Out-of-use sealed radiation sources or material used to make the sources, accelerator-produced radioactive materials
Uranium and thorium ore processing wastes	Recovery of uranium or thorium for DOE or civilian nuclear applications	Mining and milling tailings, process residues, soils, equipment
Naturally occurring and technologically enhanced naturally occurring radioactive materials (NORM and TENORM)	Recovery and processing of mineral resources unrelated to nuclear applications, municipal water treatment	Commercial ore mining residues, phosphate mining and fertilizers, scale and sludge from oil and gas production, water treatment filters, resins, and sludges

Ecology), on February 6-7, 2003, and Salt Lake City, Utah (Envirocare of Utah), on April 16-17, 2003. Four committee members visited FUSRAP[7] sites near St. Louis, Missouri, on May 12, 2003. A final information-gathering meeting was held in Washington, D.C. on June 12, 2003.

The following published studies served as cornerstones for the committee's deliberations and findings:

- *Risk-Based Classification of Radioactive and Hazardous Chemical Wastes* was published in 2002 by the National Council on Radiation Pro-

---

[7]Formerly Utilized Sites Remedial Action Program (see Chapter 3).

Reason for Category

These are the wastes that are typically described as low-level wastes and disposed in near-surface facilities at DOE or commercial sites. They meet the statutory definitions of Atomic Energy Act (AEA) 11e.(1) byproduct materials and the exclusionary definition of "low-level waste" under the Nuclear Waste Policy Act (NWPA, see Chapter 2). The relatively short-lived radioactive isotopes and activity levels of the waste fit into USNRC Classes A, B, or C (See Chapters 2 and 3).

These wastes represent the low end of the spectrum of materials defined, regulated, and managed as low-level waste. They produce very low or essentially undetectable levels of radiation, but they arise in very large volumes. The descriptor "slightly radioactive solid materials" was introduced in NRC, 2002a.

These wastes represent the high-end materials that meet the statutory definitions of "low-level waste," although they may be capable of producing acute radiation effects in humans and serious contamination incidents. They may exceed the USNRC Class C limit for waste that is acceptable for near-surface disposal.

These low-activity wastes do not meet the NWPA definition of "low-level waste," but they are federally regulated under the AEA definition 11e.(2) byproduct materials (see Chapter 2). Their radioactivity arises from long-lived, natural uranium and thorium isotopes and their decay products. There are large volumes of these wastes, some dating back to the Manhattan Project (see Chapter 3).

These low-activity wastes meet neither the NWPA definition of "low-level waste," nor the AEA definition 11e.(2) byproduct materials—hence they are not directly subject to federal regulation (see Chapter 2). The origin of the radioactivity in these wastes and their large volumes are comparable to uranium and thorium ore processing wastes.

tection and Measurements (NCRP). This report found that the existing patchwork system of regulations is inconsistent and becoming increasingly complex. It presents the NCRP's recommendations for a waste classification system that would apply to any waste containing radionuclides or hazardous chemicals (NCRP, 2002).

• *The Disposition Dilemma: Controlling the Release of Solid Materials from Nuclear Regulatory Commission-Licensed Facilities* was published in 2002 by the National Academies' Board on Energy and Environmental Systems. This study was requested by the USNRC to inform rulemaking on disposition of very-low-activity wastes, mainly steel and concrete from commercial nuclear reactor decommissioning. The study found that the USNRC's current approach of case-by-case clearance decisions was protective of public health, but inconsistently applied. The study recommended use of a dose-based standard in evaluating disposition options (NRC, 2002a).

• *Evaluation of Guidelines for Exposures to Technologically Enhanced Naturally Occurring Radioactive Materials* was published in 1999 by the National Academies' Board on Radiation Effects Research. This study was requested by the Environmental Protection Agency (EPA) and reflected the agency's awareness of the hazards of NORM and attempts to develop regulatory guidelines. The study found that differences among existing guidelines were based on policy judgments rather than on scientific information (NRC, 1999a).

• *United States of America National Report: Joint Convention on the Safety of Spent Fuel Management and on the Safety of Radioactive Waste Management* summarizes policies, practices, regulations, and inventory of all declared wastes in the United States. The report was prepared by the DOE, EPA, USNRC, and State Department to meet reporting requirements of the Joint Convention, which was ratified and signed by President Bush in April 2003 (DOE, 2003).

The committee also used information from the Manifest Information Management System that provides data on waste sent to commercial disposal facilities over past 12 years (http://mims.apps.em.doe.gov) and the Central Internet Database that provides information on DOE wastes (http://cid.em.doe.gov).

Outline of This Report

The committee itself had difficulty in comprehending the many complicated statutes and regulations that apply to low-activity wastes. The committee therefore felt it would be useful to begin this interim report by describing these statutes and regulations in Chapter 2. Chapter 3 summarizes low-activity waste inventories, hazards, and management and disposal practices according to the present regulatory system. Chapter 4 gives the committee's views and findings with illustrative examples.

Appendix A: Interim Report

2
The Statutory and Regulatory Context for Low-Activity Waste Management

From the discovery of radioactivity in 1895 through most of the first half of the 20th century, radioactive elements such as uranium and thorium used in industry and medicine in the United States were regulated by the states. In the middle of the 20th century the Army Corps of Engineers managed the first large-scale uses of radioactive materials in the Manhattan Project, which produced the world's first nuclear weapons. These activities were kept secret until after World War II.

Weapon component manufacturing along with other uses of materials controlled under the wartime program were first regulated under the Atomic Energy Act of 1946, the McMahon Act. The McMahon Act was intended to ensure the security of nuclear materials rather than to control their radiological hazards. It defined three categories of regulated radioactive material (source, byproduct, and special nuclear) that have been preserved in subsequent revisions of the Act and that are used in other laws and regulations (see Appendix C). The Act also created the Atomic Energy Commission (AEC) to oversee all nuclear activities begun in the Manhattan Project (DOE, 1996).

The material categories and definitions in the McMahon Act were established before the health hazards of nuclear radiation were fully appreciated—nuclear security was the overriding concern. Over the past 60 years, new regulations based on these original definitions developed as a patchwork while knowledge was gained, new materials and technologies discovered, and risks recognized. It is in this context that the Board on Radioactive Waste Management initiated this study and the committee developed its findings for this report.

FEDERAL STATUTES APPLICABLE TO LOW-ACTIVITY WASTES

The Atomic Energy Act of 1954 (AEA) replaced the McMahon Act, ended the government monopoly on use of nuclear materials, and established the framework for the commercial nuclear industry. In 1974, the Energy Reorganization Act disbanded the AEC and established the Nuclear Regulatory Commission (USNRC) to control commercial nuclear activities, and the Energy Research and Development Administration (ERDA) to control defense nuclear activities. The Department of Energy (DOE) replaced ERDA in 1977. The Environmental Protection Agency (EPA) was established in 1970 and has authority under the AEA to set radiation protection criteria and standards and issue radiation protection guidance for federal agencies. EPA also controls radioactive material under authorities granted by other statutes. Statutes that provide authority for the federal regulation of low-activity wastes are listed and described briefly in Sidebar 2.1.

Most low-activity wastes fall under provisions of the AEA because they arose as source, byproduct, or special nuclear materials. Notable exceptions are wastes that contain naturally occurring radioactive materials (NORM) from nonnuclear activities, such as mining, oil and gas production, and water treatment. Wastes that include NORM are federally regulated only if the waste, or the feedstock in processes that produced the waste, contains uranium or thorium in concentrations greater than 0.05 percent by weight (i.e., AEA source material).

Federal statutes define one important group of low-activity wastes—low-level wastes—only by exclusion: low-level waste is not spent nuclear fuel, high-level waste from fuel reprocessing, transuranic waste, or AEA section 11e.(2) byproduct material (waste from processing of uranium or thorium ore). Thus, at this time there is no statutory upper limit or lower limit for the level of radioactivity required to declare a material to be low-level waste.[1] As a result the radioactivity in wastes that meet the definition of low-level waste may be low enough that it is essentially undetectable or high enough to produce acute harm to humans or serious contamination incidents.

---

[1]Upper limits on the concentrations of radionuclides in low-level waste that can be disposed in near-surface facilities are imposed by the USNRC in 10 CFR Part 61. The USNRC has embarked on a rulemaking for the disposition of solid materials that contain very low levels of radioactivity.

SIDEBAR 2.1
Statutes Relevant to the Regulation and
Management of Low-Activity Waste

Atomic Energy Act of 1954, As Amended

The purpose of the Atomic Energy Act (AEA) (42 U.S.C. Sect. 2011-Sect. 2259) is to assure the proper management of source, special nuclear, and byproduct material. The AEA and the statutes that amended it delegate the control of nuclear energy primarily to the DOE, USNRC, and EPA. The AEA provides the following definitions:

- source material — (1) uranium, thorium, or any other material that is determined by the USNRC pursuant to the provisions of Section 61 of the AEA to be source material; or (2) ores containing one or more of the foregoing materials, in such concentration as the USNRC may by regulation determine from time to time (AEA, Section 11[z]);
- special nuclear material — (1) plutonium, uranium enriched in the isotope 233 or the isotope 235, and any other material that the USNRC, pursuant to the provisions of Section 51 of the AEA, determines to be special nuclear material, but does not include source material; or (2) any material artificially enriched by any of the foregoing, but does not include source material (AEA, Section 11[aa]); and
- byproduct material — (1) any radioactive material (except special nuclear material) yielded in or made radioactive by exposure to radiation incident to the process of producing or utilizing special nuclear material, and (2) the tailings or wastes produced by the extraction or concentration of uranium or thorium from any ore processed primarily for its source material content (AEA, Section 11[e]).

Byproduct material declared as waste is usually referred to as 11e.(1) or 11e.(2) waste, consistent with the AEA definitions.

The AEA references the Nuclear Waste Policy Act of 1982 (NWPA, see below) for the definition of high-level radioactive waste, spent nuclear fuel, and the exclusionary definition of low-level radioactive waste. A definition of transuranic waste (material contaminated with elements of atomic weight greater than 92) was added to the AEA in 1988.

Reorganization Plan No. 3 (1970)

Although this is not a statute, it was significant in delineating the responsibilities and interactions of the federal agencies.

continued

SIDEBAR 2.1 Continued

When the Environmental Protection Agency (EPA) was created, it received certain functions and responsibilities from other federal agencies. Among the functions transferred to EPA was the AEA authority to "establish generally applicable environmental standards for the protection of the general environment from radioactive material. As used herein, standards mean limits on radiation exposures or levels, or concentrations or quantities of radioactive material, in the general environment outside the boundaries of locations under the control of persons possessing or using radioactive material." EPA also received the functions of the Federal Radiation Council, including the responsibility to develop and issue radiation protection guidance to all federal agencies.

Energy Reorganization Act (1974)

The Energy Reorganization Act amended the AEA to split the federal authority over the defense and civilian uses of nuclear materials and facilities. The Atomic Energy Commission was replaced by two new entities. The Nuclear Regulatory Commission (USNRC) became responsible for the regulation of civilian nuclear facilities and activities, and the Energy Research and Development Administration (ERDA) became responsible for defense-related nuclear facilities and activities—including regulation of defense program wastes, and civilian nuclear research and development activities, e.g., advanced reactors.

Department of Energy Organization Act (1977)

The Department of Energy Organization Act created the DOE as a cabinet-level agency. DOE replaced ERDA, combined parts of several other agencies, and took over responsibility for defense program wastes.

Nuclear Waste Policy Act of 1982, As Amended

The NWPA provided statutory definitions for the terms "high-level radioactive waste" (HLW) and "spent nuclear fuel." However, the NWPA defined "low-level radioactive waste" (LLW) in terms of what it is not. That is, LLW is defined as material that is not HLW, spent nuclear fuel, transuranic waste, or AEA 11e.(2) byproduct material. The NWPA provides authority for the USNRC to classify material as HLW. Waste containing naturally occurring or accelerator-produced radioactive material (i.e., non-AEA-defined nuclear fuel cycle material) is not included in the NWPA.

Uranium Mill Tailings Radiation Control Act of 1978

The Uranium Mill Tailings Radiation Control Act (UMTRCA) addresses the regulation and control of uranium mill tailings (byproduct material as defined in section 11e.(2) of the AEA). UMTRCA vested the EPA with overall responsibility for establishing health and environmental cleanup standards for uranium milling sites and contaminated vicinity properties, the USNRC with responsibility for licensing and regulating uranium production and related activities including decommissioning, and the DOE with responsibility for remediation of inactive mill tailings sites and long-term monitoring of all the decommissioned sites.

Low-Level Radioactive Waste Policy Act of 1980, As Amended in 1985

The Low-Level Radioactive Waste Policy Act (LLRWPA) establishes state (including regional compacts of states) and federal responsibility for the disposal of LLW and defines the roles of federal agencies (particularly the DOE and the USNRC). The LLRWPA also refers to the USNRC classification of LLW in 10 CFR Part 61. The definition of LLW is essentially the same as in the NWPA, although transuranic wastes are not specifically excluded in the 1985 Amendments.

Comprehensive Environmental Response, Compensation, and Liability Act of 1980, As Amended by the Superfund Amendments and Reauthorization Act of 1986

The Comprehensive Environmental Response, Compensation, and Liability Act (CERCLA), also known as Superfund, gives the EPA, in conjunction with state regulators, the authority to investigate and remediate sites placed on the National Priority List. The full process includes site characterization, evaluation of alternative remediation strategies, and public involvement and results in a legal Record of Decision (ROD). Many sites contaminated with radioactive material, including those licensed by USNRC or controlled by DOE, have been placed on the National Priority List. Guidance for cleaning up contaminated soil and materials, including TENORM, have been issued by EPA.

Resource Conservation and Recovery Act of 1976

The Resource Conservation and Recovery Act (RCRA) has been amended several times, with the most significant amendments passed in 1984 as the

continued

SIDEBAR 2.1 Continued

Hazardous and Solid Waste Amendments. RCRA provides for the cradle-to-grave control of chemically hazardous wastes by imposing management requirements on generators and transporters of hazardous waste and on owners and operators of treatment, storage, and disposal facilities. Regulations pertaining to RCRA waste disposal facilities (landfills) include such details as liner and cover designs.

The RCRA hazardous waste regulations are found in Title 40 of the Code of Federal Regulations. Parts 260 through 265 describe hazardous waste management, provide EPA's lists of hazardous wastes, and set standards that must be met by hazardous waste generators and managers. EPA's land disposal restrictions are given in Part 268 and its permit programs in Part 270.

RCRA specifically excludes material regulated under the AEA from its jurisdiction; however, RCRA is applicable to the hazardous constituents in waste contaminated with both chemically hazardous and radioactive materials, which could include accelerator-produced materials.

FEDERAL REGULATIONS APPLICABLE TO COMMERCIAL LOW-ACTIVITY WASTES

At the federal level, AEA low-activity wastes generated or disposed in the commercial sector are regulated by the USNRC under its authority to license nuclear facilities and the possession of nuclear materials (see Appendix A). The USNRC may relinquish a portion of its authority to individual states, known as Agreement States. All disposal facilities currently licensed to accept low-level wastes are located in Agreement States. The EPA has authority to regulate environmental radiation exposure as well as hazardous chemical wastes, and in certain cases to determine appropriate waste disposal and cleanup methods.

Low-activity wastes that contain both AEA radionuclides and hazardous chemicals are referred to as "mixed wastes" and are thus subject to regulation by both the USNRC and EPA. The Department of Transportation regulates the shipment of radioactive materials while the USNRC has the authority to regulate certain packages for transportation of nuclear materials. Sidebar 2.2 summarizes federal regulations for low-activity wastes in the commercial sector.

Non-AEA wastes, such as TENORM wastes, are subject to EPA radiation protection standards and guidance. The Resource Conservation and

SIDEBAR 2.2
Federal Regulations That Apply to
Commercial-Sector Low-Activity Wastes

10 CFR Part 61, Licensing Requirements for Land Disposal of Radioactive Waste

These USNRC requirements apply to all LLW containing source, special nuclear, or byproduct material that are acceptable for disposal in a near-surface facility. LLW waste is defined the same way as it is defined in the LLRWPA and the NWPA, namely, radioactive waste that is not classified as high-level radioactive waste, transuranic waste, spent nuclear fuel, or byproduct material as defined in section 11e.(2) of the AEA (i.e., uranium or thorium tailings and waste). Part 61.55 defines three LLW classes (A, B, and C) that are acceptable for disposal in near-surface facilities. Greater than Class C (GTCC) low-level radioactive wastes are the responsibility of DOE. The DOE must dispose of GTCC wastes in a deep geologic disposal facility licensed for high-level waste or in some other manner approved by the USNRC. [NOTE: Federal government responsibility for GTCC is not in the regulations, but in the 1985 LLRWPA Amendments.]

10 CFR Part 20, Subpart K, Waste Disposal

This regulation addresses disposal by release into sanitary sewers, treatment or disposal by incineration, and disposal of specific wastes that are below specified activity levels.

10 CFR Part 40, Domestic Licensing of Source Material, Appendix A, Criteria Relating to the Operation of Uranium Mills and the Disposition of Tailings or Wastes Produced by the Extraction of Concentration of Source Material from Ores Processed Primarily for their Source Material Content (Incorporating 40 CFR Part 192, "Health and Environmental Protection Standards for Uranium and Thorium Mill Tailings")

The criteria apply to uranium mill tailings (section 11e.[2] material under the AEA) generated at mill sites licensed in or after 1978, the date of enactment of the Uranium Mill Tailings Radiation Control Act. Under the USNRC's interpretation of UMTRCA, the Commission does not have jurisdiction to regulate mill tailings generated prior to 1978.

continued

SIDEBAR 2.2 Continued

40 CFR Part 266, Standards for the Management of Specific Hazardous Wastes and Specific Types of Hazardous Waste Management Facilities

Subpart N of these standards exempts certain mixed waste from RCRA requirements if it satisfies specific criteria.

40 CFR Part 300, National Oil and Hazardous Substances Pollution Plan

This regulation implements CERCLA, including the identification of applicable or relevant and appropriate requirements (ARARs). ARARs are specified on a case-by-case basis in each Record of Decision (ROD). When there is no ARAR, or when the ARAR is considered to be nonprotective, a lifetime risk range of 10^{-4} to 10^{-6} is used.

2003 Advance Notice of Proposed Rulemaking (anticipated)

The EPA is requesting public comment on methods to define and alternatives for disposal of low-activity radioactive waste, including exemption for mixed wastes containing small amounts of radioactive material for disposal in a RCRA Class C disposal cell.

Recovery Act (RCRA) provides another important authority for the EPA to regulate non-AEA material. States must go through a formal delegation process to receive EPA authorization to implement the RCRA hazardous waste program, but EPA leaves implementation of RCRA solid waste provisions almost entirely to the states.[2] Radiation protection responsibilities may also be delegated to individual states. As noted later in this report, there are significant differences in the states' approaches to regulating low-activity wastes.

In addition to the primary federal regulations summarized in Sidebar 2.2, several other regulations affect the quantity and disposition of low-activity wastes. Materials that cannot be released or that are contaminated in decommissioning or site cleanup work will become waste. For example, the USNRC regulations governing the decommissioning of licensed sites contaminated with residual radioactive material establish a

---

[2]Most TENORM wastes are categorized as solid wastes but not as hazardous waste and thus are state-regulated.

25 millirem/year dose criterion for the release of a site for restricted or unrestricted use (10 CFR Part 20, Subpart E, Radiological Criteria for License Termination). Similarly the EPA has developed a 15 millirem/year criterion for the cleanup of soils contaminated with radioactive material (OSWER No. 9200.4-18 Establishment of Cleanup Levels for CERCLA Sites with Radioactive Contamination).

The EPA has exercised its authority under the Clean Air Act to develop standards that limit radon emissions from surface sources (for example, 40 CFR Part 61, Subpart R, National Emission Standards for Radon Emissions from Phosphogypsum Stacks) and subsurface natural geologic deposits on which structures are built, and radioactive emissions from DOE facilities (40 CFR Part 61, Subpart H, National Emission Standards for Emissions of Radionuclides other than Radon from Department of Energy Facilities). The EPA has the authority to regulate non-AEA radioactive waste under the Toxic Substance Control Act (TSCA—15 U.S.C. S/S 2601 et seq. 1976) but has not exercised this authority to date.

DEPARTMENT OF ENERGY CONTROL OF LOW-ACTIVITY WASTES

The manufacture of nuclear weapons, which began with the Manhattan Project, is now the responsibility of the DOE—along with responsibility for radioactive waste left as a legacy of the Cold War (DOE, 1996).[3] The DOE is self-regulating for low-level waste (LLW) generated and disposed on its own sites. To determine which wastes are deemed to be LLW, DOE uses the exclusionary definition of LLW provided by the Nuclear Waste Policy Act of 1982 (NWPA), as amended. Accordingly, DOE manages all waste as LLW unless it meets the definition of high-level waste, spent fuel, transuranic waste, or byproduct material (as defined in section 11e.[2] of the AEA, as amended). DOE excludes NORM waste from its definition of LLW, but regulates potential exposures under its radiation protection directives and often manages small amounts of NORM as LLW. LLW that contains hazardous substances as defined by the EPA in 40 CFR Parts 260 and 261 is managed as mixed low-level waste.

In addition to promulgating regulatory requirements that have the force of law, e.g., 10 CFR Part 835 (see Sidebar 2.3), DOE has developed a number of Orders addressing radioactive waste and other issues. These DOE Orders do not have the legal enforcement mechanism of a federal regulation. Instead, DOE Orders are incorporated by reference into individual government contracts, and the provisions of the referenced DOE

---

[3]The Department of Defense is responsible for U.S. military operations, including deployment of nuclear weapons.

SIDEBAR 2.3
DOE Regulations and Orders

DOE Order 435.1, Radioactive Waste Management (1999) (together with corresponding Manual (DOE M 435.1-1) and Implementation Guide (DOE G 435.1-1))

DOE Order 435.1 covers all HLW, transuranic waste, and LLW handled by all elements of DOE, including accelerator-produced waste and the radioactive component of mixed waste. It also covers both byproduct material as defined by section 11e.(2) of the AEA, as amended, and naturally occurring radioactive material when the byproduct material or naturally occurring radioactive material are managed at DOE LLW facilities. Order 435.1 does not apply to spent fuel from nuclear reactors. Chapter IV of the manual addresses LLW. DOE does not classify wastes using the USNRC's Class A, B, C system. For DOE, the location of its sites is confined to the location of its facilities, and only DOE generators send waste to them. Thus, DOE individually evaluates the performance capabilities of its sites and establishes waste acceptance criteria for each based on a site-specific assessment.

10 CFR Part 835, Occupational Radiation Protection (1998)

DOE's radiation protection requirements are equivalent to those contained in the requirements for the commercial sector in 10 CFR Part 20 and are contained in two separate directives. The first is 10 CFR Part 835, which addresses occupational radiation protection. It establishes radiation standards, limits, and program requirements for protecting individuals from ionizing radiation resulting from the conduct of DOE activities. Part 835 requires that DOE activities involving occupational radiation exposure "shall" be conducted in compliance with a documented radiation protection program (RPP) as approved by DOE. Effective occupational radiation protection programs ensure that the health and safety of the work force are

---

Orders are enforced through contract oversight. This system is complex and tends to vary from contract to contract and over time. To address this issue, DOE embarked on a program of replacing many of its Orders with regulations. However, several years ago DOE abandoned this effort as being too cumbersome.

adequately protected by maintaining individual and collective radiation doses below regulatory limits and by implementing a process that seeks doses that are as low as is reasonably achievable (ALARA). The documented RPP includes the programs, plans, procedures, schedules, and other measures undertaken to ensure worker health and safety through compliance with 10 CFR Part 835. The rule applies to exposures from the management of waste at DOE facilities and contains requirements for controlling property that may be contaminated.

DOE Order 5400.5, Radiation Protection of the Public and the Environment (1990)

DOE Order 5400.5 requires DOE facilities to maintain public doses of radiation below established limits and constraints and as low as practicable below the limits using the ALARA process. The order contains requirements for limiting liquid discharges and air emissions. It includes requirements to limit sewer discharges and use of soil columns for controlling disposed radioactive material. Order 5400.5 also contains DOE's requirements for managing technologically enhanced NORM and 11e.(2) byproduct material and DOE's process for control and release of property from DOE control. Property containing low levels of residual radioactive material may be released for unrestricted (e.g., release for residential use of a property) or in some cases, restricted use (e.g., disposition of waste or other personal property to a RCRA landfill or release of real property for recreational use only) if the levels are shown to be below DOE-approved authorized limits. Property demonstrated to meet surface activity guidelines may be released for unrestricted use. Alternatively, unrestricted release or restricted release may be done to authorized or supplemental limits developed and approved (by DOE) on a case-by-case basis if they meet dose constraints and ALARA process requirements.

STATE REGULATIONS

Federal statutes provide three important responsibilities for the states with regard to low-activity wastes: (1) each state must have a way to dispose of its own LLW (but not NORM wastes); (2) states may assume portions of the USNRC's regulatory authority by becoming an Agreement State for the regulation of LLW or uranium mill tailings; and (3) the states regulate non-AEA wastes under authority provided by the state legislature (because they are not covered by federal statutes).

As noted in Sidebar 2.1, the LLRWPA of 1980 required every state to provide for disposal of its own LLW, either alone or in cooperation with other states. The law was intended to encourage the formation of regional interstate compacts, which would be ratified by Congress, for disposing of LLW. In 1985, because no compacts had been ratified or disposal sites selected, Congress amended the LLRWPA to create milestones and incentives for siting disposal facilities (see Sidebar 2.4). Although the milestones have generally been missed (only three disposal sites are operating, as will be discussed in Chapter 3), the states have formed 10 compacts, most states are members of a compact, but no new sites have been developed by the compacts. The compacts and their membership are summarized in Table 2.1.

Section 274 of the AEA, as amended, provides the statutory basis for Agreement States. The USNRC may relinquish to the states portions of its regulatory authority to license and regulate byproduct materials, source materials, and certain quantities of special nuclear materials. The mechanism for the transfer of USNRC's authority to a state is an agreement signed by the governor of the state and the chairman of the Commission.

In order for an Agreement State to license an LLW disposal facility, the state regulations for LLW disposal must be compatible with USNRC's regulations in 10 CFR Part 61. The USNRC also conducts periodic reviews of Agreement State programs, as part of its Integrated Materials Performance Evaluation Program, to determine if the state's regulations and practices continue to be adequate and compatible with USNRC's. If requested, USNRC provides assistance to the Agreement States on LLW disposal issues. Presently there are 33 Agreement States, including the three states that currently have licensed LLW disposal facilities. Several other states are in the process of reaching agreement with USNRC.

There are differences among the states as to what materials are regulated as TENORM and how they are regulated. While a few states have begun to establish a licensing system for all industries that generate TENORM wastes (similar to the way the USNRC licenses facilities that handle radioactive sources), others control this class of wastes using specific regulations for TENORM. The majority treat the waste in accordance with general radiation protection requirements. The environmental, radiation protection, and waste disposal methods in most cases are based on EPA and or USNRC regulations or guidance.[4]

An effort has been undertaken by the Conference of Radiation Con-

---

[4]The NORM Technology Connection maintained by the Interstate Oil and Gas Compact Commission provides state-specific regulatory requirements applicable to NORM-containing waste <http://www.iogcc.state.ok.us/norm/reg/state>.

SIDEBAR 2.4
Effects of the Low-Level Radioactive Waste Policy Act

In 1980, Congress enacted the LLRWPA, reflecting its declared policy of holding each state responsible for providing capacity for disposal of its LLW either within its own boundaries or through state compacts. However, Congress provided no penalties if states failed to provide disposal capacity. Five years later, there were still no assured disposal locations for such waste for at least thirty states.

In response to this failure of the majority of states to designate disposal sites within their respective borders or to enter into regional compacts, Congress again addressed this disposal issue in the Low-Level Waste Policy Amendments Act of 1985. To stimulate the states into action, Congress provided three types of incentives. The first was to provide those states that did enter into regional compacts with monetary incentives. The second was to allow states hosting disposal sites to impose substantial surcharges for waste disposal on those states that failed to comply, and, after 1990, to deny noncompliant states access to disposal facilities. The third incentive provided that if a state was unable to provide for disposal of its waste by 1996, then the state could be required to take title of the waste from the waste generator and take possession of the waste. In addition, the state would be liable for any damages incurred by the waste generator resulting from the failure of the state to take the waste.

In 1990, the State of New York filed suit claiming that the 1985 amendments were in violation of various provisions of the Constitution. Although the State of New York initially lost the case, U.S. Supreme Court agreed to hear the State's case on appeal and ultimately ruled in favor of the State on some of the issues raised (*State of New York v. United States*, 488 U.S. 1041 (1992)).

The Court noted that although Congress may encourage, or provide incentives for, states to regulate in a particular way, it could not coerce a state into action. The Court went on to find the first and second incentives provided in the 1985 amendments to be permissible under the Constitution. However, the Court also found the third incentive to be constitutionally prohibited coercion in which Congress attempted to compel the states to regulate LLW disposal. Thus, the Court struck down the third incentive, while allowing the other two to remain intact to encourage state action.

The Court concluded that although the third incentive was prohibited, Congress nevertheless might have many other methods of achieving its goal of regional self-sufficiency in LLW disposal. However, in more than a decade since the Court's ruling, Congress has not revisited this issue.

TABLE 2.1 Interstate Compacts for Low-Level Waste Disposal

Compact Name	Associated States
Northwest	Alaska, Hawaii, Idaho, Montana, Oregon, Utah, Washington, Wyoming
Southwestern	Arizona, California, North Dakota, South Dakota
Rocky Mountain	Colorado, New Mexico, Nevada
Midwest	Indiana, Iowa, Minnesota, Missouri, Ohio, Wisconsin
Central	Arkansas, Kansas, Louisiana, Nebraska, Oklahoma
Texas	Maine, Texas, Vermont
Central Midwest	Illinois, Kentucky
Appalachian	Delaware, Maryland, Pennsylvania, West Virginia
Atlantic	Connecticut, New Jersey, South Carolina
Southeast	Alabama, Florida, Georgia, Mississippi, Tennessee, Virginia
Unaffiliated States	District of Columbia, Massachusetts, Michigan, New Hampshire, New York , Puerto Rico, Rhode Island, North Carolina

SOURCE: USNRC, 2002.

trol Program Directors (CRCPD) to develop a model TENORM regulation that could be adopted or modified by state agencies for use in their particular state. The model regulation[5] (Suggested State Regulations for Control of Radiation—Part N) would require licensing of companies which possess, use, manufacture, or make products or wastes in which the radium-226 content is $\geq$ 5 picocuries/gram. As of this writing, the model regulation has been redrafted a number of times. Once the draft regulation is approved by the CRCPD board of directors, it will be provided to several federal agencies (including EPA, USNRC, and DOE) for their comments and concurrence. If approved, the regulation would be published for states to consider in developing their own approaches to TENORM.

EVOLUTION OF THE RISK CONCEPT FOR CONTROLLING LOW-ACTIVITY WASTES

Risk does not explicitly appear in current statutes or regulations that control low-activity waste; rather risk is an evolving concept that is receiving increased attention by policy makers, regulators, and members of the public. This section provides a brief history of the concept's initial development from radiation dose-based regulations. In its final report the

---

[5]See <http://crcpd.org/SSRCRs/TOC_8-2001.htm>.

committee will address the concept of risk and options for using risk to better inform future regulatory policies for low-activity wastes.

As noted earlier in this chapter, the Atomic Energy Act of 1946 (McMahon Act) was intended to ensure security of nuclear materials rather than to control their hazards to workers or the public. The earliest controls for releases of radioactive materials from licensed activities, in air or water effluents, were set by the AEC in 10 CFR Part 20. These control levels for individual radioisotopes were set with the idea of controlling the exposure of the persons closest to the site, based on directly measurable effluents at the site boundary for liquid effluents or the point of release for gaseous effluents.

International principles for radiation protection were adopted as part of applying the effluent limits, including the ALARA principle. This principle is followed when radioactive releases are not only controlled to strict limits, but are also controlled so that releases, or exposures, are "as low as reasonably achievable" (ALARA). The ALARA principle was applied to effluent control, e.g., to nuclear reactor gaseous effluents through 10 CFR Part 50, Appendix I (1975).

Years later, when the EPA developed new emission limits for radionuclides under the Clean Air Act (NESHAPS), 40 CFR Part 61, they were based directly on 10^{-4} (one chance in 10,000) lifetime risk of cancer death, corresponding to an exposure of about 10 mrem/yr to the maximally exposed individual. In retrospect, the EPA concluded that the USNRC programs for fuel cycle facilities, including 10 CFR Part 50, Appendix I, for reactors, provided adequate risk protection and amended the NESHAPS accordingly.[6]

In the early 1980s the USNRC developed an Environmental Impact Statement (EIS) for a typical shallow land disposal site for LLW (NUREG-0945, Final Environmental Impact Statement on 10 CFR Part 61, "Licensing Requirements for Land Disposal of Radioactive Waste," USNRC November 1982). In this EIS the requirements for licensing LLW disposal were

---

[6]EPA's policy is to apply a consistent risk management approach to all of its programs and statutory mandates. CERCLA regulations call for cleanups to achieve a residual lifetime risk of between 1 in 1,000,000 [10^{-6}] and 1 in 10,000 [10^{-4}] (40 CFR 300.430(e)(2)(i)(A)(2)). When applied to radiation, EPA considers a dose of 15 mrem/yr over a lifetime to correlate to a risk of approximately 3×10^{-4} (3 in 10,000), which is considered "essentially equivalent" to the 1×10^{-4} target (OSWER directive 9200.4-18, August 22, 1997).

Following the CERCLA approach, EPA explicitly considers risk implications in other actions involving radiation. In 1989, EPA established airborne emission limits for a wide variety of source categories under the Clean Air Act (NESHAPs), 40 CFR Part 61. EPA's approach to establishing limits required first that an "acceptable risk" level be established with a presumptive limit on maximum individual risk of approximately 1 in 10,000.

developed by analyzing the potential releases from a large burial site containing typical amounts of various forms of LLW, given imposition of the licensing requirements being considered. The measure of impact was not risk directly, but radiation dose to persons near the site boundary, analyzed to occur at any time far into the future. This same dose-basis analysis has been adopted by DOE in the Order 435.1 guidance.

Appendix A: Interim Report

3
Low-Activity Waste Overview

This chapter summarizes current low-activity radioactive waste (LAW) management and regulatory practices in the United States. The first section provides information on the characteristics, inventories, and regulatory controls for wastes in each of the categories introduced in Chapter 1. The second section provides a perspective on the radiological hazards of these wastes. The final section describes currently available disposal sites and disposal practices. In developing this chapter the committee has focused on the relevant information that led to its findings, rather than reproducing the detailed summary information available elsewhere.[1]

Among the wastes described in this chapter, low-level wastes (LLW) from Department of Energy (DOE) and commercial nuclear facilities have received the most attention from regulators and the public.[2] LLW in the form of debris, rubble, and contaminated soils from facility decommissioning and site cleanup constitute much larger volumes than LLW from operational facilities but generally contain very low concentrations of radioactive material. Discrete radioactive sources that are no longer use-

---

[1]Detailed summary information is available from DOE (1999, 2001, 2003), the Manifest Information Management System (MIMS) at <http://mims.apps.em.doe.gov>, and the Central Internet Database <http://cid.em.doe.gov>. Note that DOE (1999, 2003), and MIMS provide commercial-sector data.

[2]LLW fall under the Atomic Energy Act. They are defined in the Nuclear Waste Policy Act of 1982 by exclusion, namely waste that is not spent fuel, high-level waste from fuel reprocessing, transuranic waste, or 11e.(2) byproduct material (see Chapter 2).

ful also meet the definition of LLW even though they may contain highly concentrated radioactive material. Although similar in their characteristics, DOE "defense" LLW and commercial LLW are generally managed and regulated separately according to their respective origins in the DOE or private sector.

Tailings and other wastes from mining and processing uranium and thorium ores have been produced in very large quantities. Like LLW, uranium and thorium wastes are subject to the Atomic Energy Act (AEA), but concern about them has been limited mainly to populations living around mining and milling sites—including Native Americans. Non-nuclear enterprises such as mineral recovery and water treatment produce equally large or larger volumes of wastes that contain the same naturally occurring radioactive materials (NORM) as uranium and thorium wastes. NORM wastes are not subject to the AEA, and there is no consistent system for regulating them.

COMMERCIAL LOW-LEVEL WASTE

Commercial LLW comes from nuclear power facilities and other industrial, medical, and research applications. Typical examples include protective shoe coverings and clothing, mops, rags, equipment and tools, laboratory apparatus, process equipment, reactor water treatment residues, non-fuel-bearing hardware, and some decontamination and decommissioning wastes. LLW are produced in essentially every state. With a few exceptions, the radionuclides contained in commercial LLW are relatively short-lived fission products.

The 1978 revision of the AEA gave the Nuclear Regulatory Commission (USNRC) authority to regulate wastes from the private sector. Defense LLW becomes subject to USNRC regulations if it is shipped for disposal in a commercial facility. In its regulations governing the disposal of commercial LLW, the USNRC defines three classes (A—the least hazardous—B, and C) based largely on the concentrations and half-lives of radionuclides in the waste. High or essentially unrestricted concentrations of radionuclides with half-lives less than 5 years are allowed, concentrations of some specific fission and activation products with longer half-lives are restricted, and concentrations of transuranic nuclides with half-lives greater than 5 years are limited to 100 nanocuries/gram (see Appendix A, Tables B.1 and B.2). The vast majority of the volume of commercial LLW consists of the least hazardous USNRC Class A waste.

The Manifest Information Management System (MIMS) provides information on waste shipments to commercial disposal facilities (Barnwell, South Carolina; Clive, Utah; and Richland, Washington, discussed later in this chapter).[3] According to MIMS, approximately 600,000

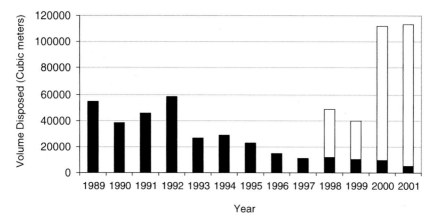

FIGURE 3.1 Volumes of low-level waste disposed at commercial sites. Upper bars beginning in 1998 are very low level wastes received at Envirocare of Utah. SOURCE: MIMS, 2003.

cubic meters of waste containing almost 9 million curies of radioactivity were disposed from 1989 through 2001 (see Figures 3.1 and 3.2). The vast majority of the waste, some 85 percent of the volume and the curies, came from nuclear utilities. Wastes from other industries amounted to about 7 percent of the volume and the curies. Wastes received from DOE sites made up most of the remainder. Waste from medical and academic origins amounted to less than 1 percent of the volumes and curies disposed.

The trend toward volume reduction begun in the mid-1990s resulted from significant efforts to reduce waste production and to further reduce volume by compaction and supercompaction of waste. The substantial volume increase beginning in 1998 is the result of large amounts of slightly contaminated soils, debris, and rubble that Envirocare of Utah began receiving in that year. The waste sent to Envirocare, however, contained less that 1 percent of the curies disposed.

DOE DEFENSE LOW-LEVEL WASTE

Defense LLW has been generated in the course of producing or using special nuclear materials throughout the DOE complex, including fuel fabrication, reactor operation, and isotope separation and enrichment, and

---

[3]See <http://mims.apps.em.doe.gov>. DOE does not assure the quality of this information.

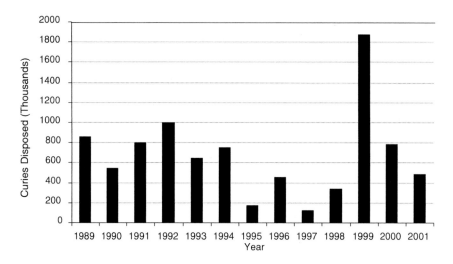

FIGURE 3.2 Curies of low-level waste disposed at commercial sites.
SOURCE: MIMS, 2003.

it continues to be produced in site cleanup work.[4] In general terms, DOE LLW is quite similar to commercial LLW except that some radionuclides specific to nuclear fuel reprocessing appear in higher quantities. For example, some DOE LLW contains transuranic isotopes, mainly plutonium, at concentrations between 10 and 100 nanocuries per gram (nCi/g).

Cumulatively through fiscal year (FY) 1999, DOE had disposed an estimated total volume of 5.8 million cubic meters of LLW and contaminated media containing almost 50 million curies. In FY 2000, DOE treated about 833,000 cubic meters of LLW and disposed about 40,000 cubic meters. DOE disposed of another 29,000 cubic meters in commercial facilities. The treated and subsequently disposed waste volumes were about equal to new additions, so the beginning and year-end inventory remained almost constant at about 146,000 cubic meters. DOE estimates that another 2 million cubic meters will be treated and disposed by 2070 (DOE, 2001; CID, 2003). DOE's main sites that generate and dispose of LLW are shown in Figure 3.3.

As noted in Chapter 2, DOE is self-regulating for wastes generated

---

[4]Department of Defense LAW is not discussed in this report. This waste is managed and disposed by contractors as commercial waste regulated by the USNRC unless it is classified for security purposes. Classified waste is managed and disposed by DOE.

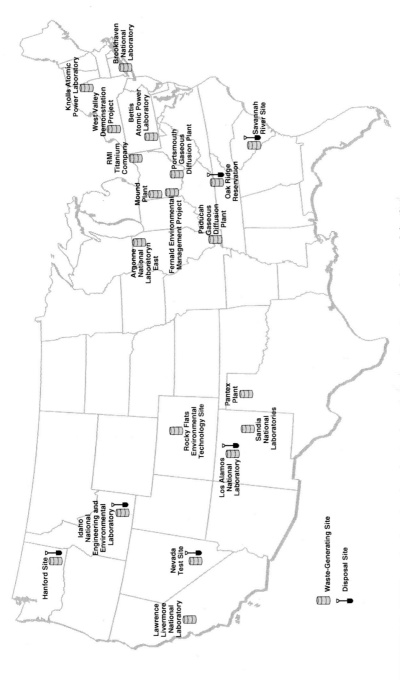

FIGURE 3.3 DOE's major low-level and mixed waste-generating sites and disposal facilities. SOURCE: GAO, 2000.

and disposed at its sites. On-site wastes that do not fit into other waste categories defined by Order 435.1 are managed and disposed as LLW. DOE LLW shipped to commercial facilities is subject to the USNRC's or the Agreement State's commercial waste regulations.

SLIGHTLY RADIOACTIVE SOLID MATERIALS

Nuclear facility decommissioning produces debris, rubble, and contaminated soil characterized by large volumes of materials having small quantities of radioactive contamination—including concrete, plastics, metals and other building materials, equipment, and packaging. A previous study (NRC, 2002a) introduced the term "slightly radioactive solid materials" (SRSM) to describe these materials. These wastes are produced in both the DOE and commercial sectors.

Decommissioning the existing commercial power reactor facilities may generate up to about 8 million cubic meters of SRSM, about 90 percent being concrete. These same facilities may also yield about a million metric tons of metallic SRSM (NRC, 2002a). DOE estimates that about 700 of its reactor and processing facilities will be fully decommissioned in the course of site cleanup (NRC, 1998). DOE also estimates that about 821,000 cubic meters of solid contaminated media may be excavated during its site cleanup activities between 2000 and 2010 (DOE, 2001).

Currently these wastes are regulated and disposed as USNRC Class A wastes, which means they must be disposed in USNRC licensed facilities (or their equivalent at DOE sites). However, these wastes usually contain very small amounts of radioactivity. Debris and rubble sent to Envirocare amounted to about 90 percent of the total LLW volume disposed in 2000, but amounted to only about 1 percent of the radioactivity (MIMS, 2003). The USNRC and its Agreement States have allowed alternative disposal pathways (e.g., in permitted landfills) on a case-by-case basis (USNRC, 2002). Both the Environmental Protection Agency (EPA) and USNRC are investigating alternative disposition options for these wastes.

DISCRETE RADIATION SOURCES

Discrete radiation sources usually consist of a radioactive material in a leak-tight metal casing. The amount and type of radioactive material used (e.g., Co-60, Sr-90, Cs-137, Ir-192, Cf-252, Am-241) determine the type and intensity of emitted radiation. Sealed sources have essential uses in medical diagnostics and therapy, industry (radiography, well logging), and research. Over the course of time, radioactive decay may reduce their intensity below a useful level, or the application may become obsolete—such as the use of Ra-226 in medicine or Cs-137 irradiators. Unused radioactive sources are often referred to as "spent" sealed sources although they may continue to present a significant radiation hazard if not properly stored or disposed (IAEA, 2001).

Sealed sources in commercial use are licensed by the USNRC or an Agreement State. DOE controls sealed sources used at its sites. As a practical matter, however, the identifying marks and records on many sealed sources, especially older sources, are sometimes lost and the sources themselves may become lost or "orphaned." According to some estimates there are over 30,000 orphan sources in the United States. In cooperation with the Conference of Radiation Control Program Directors, the EPA, USNRC, and DOE are funding a program to assist states to retrieve and securely dispose of orphan sources.[5]

While many discrete sources clearly are not low-activity materials, they meet the Nuclear Waste Policy Act definition of LLW (see Chapter 2). Their designation as LLW generally works in practice because the radionuclides in these sources typically have half-lives of a few decades or less,[6] and their small volume allows them to be safely stored in shielded containers. Regulatory authorities in most countries allow their disposal in near-surface facilities designed for LLW. Nonetheless, these sources represent the opposite extreme from the large volumes and low activities that characterize most other wastes considered in this report.

URANIUM MINING AND PROCESSING WASTES

Beginning with the Manhattan Project in 1942, uranium and thorium ores were mined and processed on a massive industrial scale (DOE, 1996). Initial ore production was dedicated to the manufacture of material for nuclear weapons; subsequent production supported the nuclear power industry as well. From the earliest days of the weapons program into the Cold War period, the government and its contractors, while maintaining the urgent pace of the program, developed an irregular pattern of waste retention and storage. The residues from recovering and processing uranium and thorium were stored in outdoor piles for later management or sometimes buried on site. In some cases tailings have been used inappropriately as construction materials (NRC, 1986).

The radiological hazards of these wastes arise from decay of naturally occurring uranium and thorium isotopes and their daughter isotopes (see Table 3.1). Beginning with Th-232, U-238, or U-235, radioactive decay

---

[5]See <http://www.epa.gov/radiation/cleanmetals/orphan.htm> and <http://www.crcpd.org/PDF/Announcement.pdf>.

[6]Radium-226 and Americium-241 are notable exceptions with half-lives of about 1600 and 460 years, respectively.

TABLE 3.1 Uranium, Thorium, and Their Longer-Lived Radioactive
Decay Products

Isotope	Half-life	Isotope	Half-life	Isotope	Half-life
U-238	4.47×10^9 y	U-235	7.04×10^8 y	Th-232	1.41×10^{10} y
Th-234	24.1 d	Pa-231	3.28×10^4 y	Ra-228	5.75 y
U-234	2.46×10^5 y	Ac-227	21.77 y	Th-228	1.91 y
Th-230	7.54×10^4 y	Ra-223	11.44 d	Pb-208	stable
Ra-226	1600 y	Pb-207	stable		
Rn-222	3.82 d				
Pb-210	22.3 y				
Po-210	138.4 d				
Pb-206	stable				

NOTE: y = years; d = days

SOURCE: NRC, 1999a.

produces a series of other radioisotopes (daughters) leading to the eventual formation of stable (nonradioactive) isotopes. The half-lives of the thorium and uranium parent isotopes are extremely long—comparable to the age of the Earth, which is why they still exist in nature. The radioactivity associated with wastes containing these isotopes is therefore low but persistent. Radon-222, a daughter product of U-238 is of particular concern because it is gaseous and can diffuse from tailings piles unless they are properly capped.

Uranium and thorium processing tailings wastes are defined as byproduct material in section 11e.(2) of the AEA (see Chapter 2). Typical tailings piles range in size from tens of thousands to over three million cubic meters (DOE, 2003). If these wastes were generated at facilities under license by the USNRC in 1978 or thereafter, they are managed under the Uranium Mill Tailings Radiation Control Act (UMTRCA) of 1978. Both the EPA and the USNRC regulate aspects of UMTRCA site remediation and waste disposal.

The USNRC has determined that it does not have authority to regulate uranium mining and processing wastes at facilities that were not under USNRC license at the time of passage of UMTRCA. Some of these wastes, generated between the start of the Manhattan Project and 1978 and related to the nation's early atomic weapons program, are managed under the Formerly Used Sites Remediation Action Program (FUSRAP) established under the AEA. FUSRAP cleanups are conducted by the Army Corps of Engineers (see Sidebar 3.1). The DOE manages uranium-contaminated wastes on its sites.

SIDEBAR 3.1
FUSRAP and UMTRCA: Two Programs for the Same Materials

The Formerly Utilized Sites Remedial Action Program (FUSRAP) is an environmental program established in March 1974 by the Atomic Energy Commission under the authority of the Atomic Energy Act of 1954. The program was created to identify, investigate, and take appropriate cleanup action at sites with radioactive contamination resulting from the nation's early atomic weapons program. Cleanup at FUSRAP sites primarily involves building debris and soils contaminated with uranium and thorium.

The DOE assumed responsibility for FUSRAP in 1977. Initially records were reviewed and surveys were performed on more than 400 sites connected with the atomic weapons program. The DOE began limited cleanups of some sites in 1979 and started major remedial actions in 1981; cleanup of 25 sites was completed by 1997.

Congress transferred responsibility for the administration and execution of FUSRAP to the Army Corps of Engineers as part of the Energy and Water Development Appropriations Act of 1998. While the Corps was assigned the responsibility for the 21 sites in the program at the time of the transfer, the DOE continues to determine the eligibility of new sites for the program. The Corps conducts cleanups under the framework of the Comprehensive Environmental Response, Compensation, and Liability Act of 1980 (CERCLA), as amended.

The Uranium Mill Tailings Radiation Control Act (UMTRCA) for uranium- and thorium-contaminated wastes was enacted in 1978. Title I of UMTRCA deals with DOE remedial action programs at former mill tailings sites, and Title II deals with non-DOE mill tailings sites and uranium mining sites that are licensed by the USNRC or an Agreement State according to USNRC regulations (see Table 2.1 in Chapter 2 for details on UMTRCA).

With FUSRAP and UMTRCA, wastes with similar radiological hazards arising mostly from uranium, thorium, and their radioactive decay products fall into different regulatory and management boxes depending on whether the materials were generated at facilities that were under license by the USNRC at the time of passage of UMTRCA in 1978. This statutory construct has led to a novel approach to managing pre-1978 ore processing residuals within FUSRAP. If the USNRC approves materials from a FUSRAP site as alternate feed material to be processed at a uranium mill for further extraction of uranium, albeit uneconomically, the residues fall under UMTRCA (because they arose after 1978) and can be put in the mill's tailings pile after processing. Some refer to this as "sham processing," an act to reclassify the waste for disposal—although from a technical standpoint the FUSRAP waste may in fact be the same as the tailings waste and the USNRC has ruled that economics is not a factor in approving alternate feed material. However, if the FUSRAP waste (or other material) is not 11e.(2) in the clear sense of the AEA, then there are significant administrative hurdles in the way of direct disposal of this material into the tailings impoundment of an UMTRCA facility.

NORM AND TENORM WASTES

Naturally occurring radioactive materials (NORM) arise in many mineral extraction operations and are often discarded as wastes—examples include phosphate industry residues, scale and sludge from oil and gas production, non-uranium mining tailings, and coal ash residues (see Table 3.2). The materials are referred to as technologically enhanced NORM (TENORM) if their concentrations of radioactive materials are increased above naturally occurring levels. Sludge or filter media from water and wastewater treatment are good examples of TENORM waste. Estimates of the NORM and TENORM inventories from U.S. industries exceed 60 billion tons (NRC, 1999a).

The radionuclides in NORM waste arise mainly from uranium and thorium series isotopes (see Table 3.1). NORM waste is therefore radiologically similar to uranium mining and milling wastes, although some radioisotope concentrations may differ. Unlike uranium and thorium wastes, NORM is not a byproduct of the production of fissionable materials and is not controlled by the AEA. Except for Department of Transportation regulations on transportation of radioactive materials, for the most part NORM is not regulated by federal agencies but rather by states.[7]

As noted in Chapter 2, there is considerable variation among states, which often regulate non-AEA materials collectively as "NARM" (see Sidebar 3.2). In Agreement States the same state agencies that have authority for AEA materials usually regulate NORM materials as well. States that regulate NORM specify concentrations of radium below which materials are exempt from regulation as waste, but the concentrations vary from state to state. Recognizing these disparities, the Conference of Radiation Control Program Directors has developed suggested state regulations for TENORM.[8]

HAZARD CONSIDERATIONS FOR LOW-ACTIVITY WASTE

The radiological hazards of LAW depend on both its level of radioactivity and its longevity. As noted by the Board on Radioactive Waste Management at the outset of this study (see Chapter 1), the radiological hazard of LAW is typically much less than that for spent nuclear fuel or

---

[7]If sites containing NORM are listed on the National Priorities List they are subject to CERCLA, and the management of the NORM wastes generated at the site are governed by applicable or relevant and appropriate requirements (ARARs), which are specified on a case-by-case basis in each Record of Decision. When there is no ARAR or when the ARAR is considered to be nonprotective, a lifetime risk range of 10^{-4} to 10^{-6} is used to establish the standard.

[8]See <http://www.crcpd.org/SSRCRs/N_4-99.PDF>.

TABLE 3.2 Domestic Processes that Generate NORM Waste

Process	Waste description	Radionuclide concentration (picocuries per gram)	Estimated waste generation (million metric tons per year)	Major generator locations
Soils in the United States	(Benchmark for typical background)	0.2 – 4.2		
Coal combustion	Fly ash	2 – 9.7	44	Midwestern and South Atlantic states
	Bottom ash and slag	1.6 – 7.7	17	
Geothermal energy production	Solids	10 – 250	0.05	California
Metal mining and processing	Slag, leachate and tailings from:			Mostly Midwestern and Western states
	–Large-volume industries[a]	0.7 – 83	1000	
	–Special application metals	3.9 – 45	0.47	
	–Rare earth metals	5.7 – 3,200	0.002	
Municipal waste treatment	Sludge[b]	1.3 – 11,600 (picocuries per liter)	3	All, especially North Central and Atlantic Coastal Plain
Oil and natural gas production	Scale and sludge	Background to over 100,000	2.6	States where petroleum or natural gas is produced or processed
Phosphate mining and fertilizer production[c]	Ore tailings and phosphogypsum (calcium sulfate)	7 – 55	48	Florida, Idaho, and other states in the West and Southeast

[a]Such as iron and copper mining.
[b]Filters typically have concentrations of 40,000 picocuries/gram but arise in much smaller volumes.
[c]Phosphate fertilizer volumes are about one order of magnitude less, with the same concentrations of radionuclides.

SOURCES: DOE, 1997, and <http://www.tenorm.com>.

SIDEBAR 3.2
NARM, NORM, and TENORM

These acronyms refer to an assortment of materials that are not subject to federal regulation under the AEA, and thus are regulated by the individual states. In many state regulations and elsewhere (e.g., NCRP, 2002) they are referred to collectively as NARM (*naturally occurring and accelerator-produced radioactive materials*).

Particle accelerators are often used to produce isotopes for medical and research purposes. In addition to these products, components of the accelerator itself may become radioactive. According to the EPA there are no firm estimates of the amount of accelerator-produced wastes, but it is generally accepted that the volume of these wastes containing isotopes with half-lives greater than one year (i.e., long enough to present waste management challenges) is very small compared to other low-activity wastes. The committee paid little attention to these materials. For completeness, however, concentrated materials with longer half-lives, e.g., Co-60, Ir-192, can be included as discrete sources in the committee's categorization of LAW. Otherwise the waste will be radiologically similar to defense or commercial LLW.

Naturally occurring radioactive materials (NORM) are a subset of NARM. They contain radioactive elements such as uranium and thorium, which were present when the Earth was formed, their radioactive decay products,[a] and some isotopes that are produced by cosmic rays from the sun—such as C-14. In its categorization the committee chose to distinguish wastes in which NORM is coincidental to recovery of mineral resources (mining, oil, gas) from wastes produced in recovery of uranium and thorium for nuclear purposes. Uranium and thorium mining and processing wastes are covered by the AEA.

Most mineral recovery operations tend to concentrate NORM to produce TENORM—technologically enhanced NORM. Examples are pipe scale, tailings piles, sludges, and filters. Water purification and treatment also produce TENORM. While noting that EPA and state regulations generally address TENORM only, the committee included both NORM and TENORM together in one category.

---

[a]Radium-226, a radioactive decay product of U-238 (see Table 3.1), was formerly used as a radiation source for medical treatments and also as a luminous paint for instrumentation. Radium effects among workers helped lead to recognition of radiation hazards. Radium wastes are no longer considered a serious issue in the United States.

high-level reprocessing waste, but the hazard may persist for very long periods. Chapter 4 will summarize the committee's view of these risks and where they fall within the current regulatory scheme. While the regulatory system was developed primarily to control radiological risks of LAW—the focus of this report—nonradiological hazards are also important.

The radioactivity in any material depends on the concentration of radioactive atoms present and their half-lives (see Sidebar 3.3). LAW are often only slightly contaminated so the radioactivity is very low. However, LAW may contain a substantial concentration of radionuclides with very long half-lives (e.g., uranium and thorium wastes, NORM wastes). The radioactivity is low, but the hazard does not diminish appreciably with time. In addition, DOE and USNRC regulations allow some wastes with relatively high radioactivity to be managed and disposed as LLW. These wastes contain fission or activation products with relatively short

---

SIDEBAR 3.3
Radioactivity in Low-Activity Wastes

The radioactivity in any material is proportional to the concentration of radioactive atoms of a given type divided by their half-life:

$$A = k \, N / t_{1/2}$$

where A is the number of radioactive disintegrations in a given time—typically disintegrations per second (becquerels) or a much larger unit (curies), equal to about 3.7×10^{10} becquerels; N is the number of radioactive atoms of a given kind (radionuclides) often expressed in units of concentration (e.g., per unit mass or volume of waste); $t_{1/2}$ is the time required for half of the initial number of radionuclides to decay (half-life); and k is a constant equal to about 0.7.

Wastes are usually contaminated with more than one radionuclide, so the total radioactivity is the sum of their individual radioactivities. The radioactivity in wastes is typically measured or calculated on the basis of volume (e.g., becquerels per cubic meter).

For slightly contaminated wastes (protective clothing, building debris, rubble) the number—or concentration—of radioactive atoms, N, is relatively small so the activity, A, is small, according to the above equation. Conversely, wastes may contain a relatively large number of radionuclides with long half-lives (uranium residues, NORM). For these wastes the quotient $(N / t_{1/2})$ is small and the radioactivity, A, is still low—but it persists for a very long time.

half-lives so their radioactivity diminishes rather rapidly—over time scales of decades to centuries.

LAW that contain chemically hazardous substances are subject to regulations of the EPA under the Resource Conservation and Recovery Act and other statutes described in Chapter 2. For these "mixed wastes," regulations of the DOE, USNRC, or Agreement States control the radioactive constituents, and EPA regulations or state permits control the chemical constituents. Chemical hazards and their regulation are described in other reports (NCRP, 2002; NRC, 1999a,b, 2002c). EPA regulations on the chemical components of mixed wastes are generally prescriptive: The Agency defines certain materials as hazardous, specifies treatment standards to be met prior to disposal, and specifies standards for construction and operation of hazardous waste sites. Institutional control, rather than site performance criteria, ensures that disposed waste remains safe.

Shipments of LAW, including NORM, are controlled by the Department of Transportation. Transportation hazards are not as well recognized as chemical hazards for LAW. Present requirements placed on waste generators along with the limited number of disposal sites result in transporting large amounts of LAW over long distances.

Envirocare of Utah receives very large amounts of slightly contaminated wastes shipped by rail and truck from all parts of the country. Plans are under way to ship the San Onofre, California, reactor pressure vessel to Barnwell, South Carolina—possibly by sea around South America because the vessel and shipping cask are too large for cross-country rail shipment and too heavy to go through the Panama Canal (St. Onge, 2003). Barnwell is the only disposal facility that can accept Class B or C waste from California (see the following section on disposal).

LOW-ACTIVITY WASTE DISPOSAL

DOE practices on-site treatment and disposal for much of the LAW generated at its major sites, which are depicted in Figure 3.3. Disposal capacity at DOE sites, especially at the Nevada Test Site and Hanford, Washington, appears to be more than adequate for future disposal needs (GAO, 2000). Nevertheless, DOE does make use of commercial treatment and disposal capabilities (described below), when appropriate for cost reduction or to supplement DOE's capabilities.

In the commercial sector, there are three sites available for disposal of LAW: Barnwell, South Carolina, operated by Chem-Nuclear; Clive, Utah, operated by Envirocare of Utah; and within the DOE Hanford site near Richland, Washington, operated by U.S. Ecology. A fourth facility at Grand View, Idaho, operated by U.S. Ecology and designed for chemically hazardous wastes, is currently receiving FUSRAP waste. Each of these facilities is limited in the types and volumes of waste that can be disposed. Sidebar 3.4 summarizes commercial waste disposal regulations and practices.

Only one disposal facility, at Barnwell, is currently accepting USNRC Class A, B, and C LLW from all states. South Carolina formed the Atlantic Compact (formerly the Northeast Compact) with Connecticut and New Jersey on July 1, 2000. Under the Compact, South Carolina can limit the use of the Barnwell facility to the three compact members. A state law enacted in June 2000 phases out acceptance of noncompact waste after 2008.

The other existing disposal facility for all three major classes of low-level waste is the Hanford, Washington, site operated by US Ecology. Controlled by the Northwest Compact, the Hanford site will continue taking waste from the neighboring Rocky Mountain Compact (see Table 2.1) under a contract.

The Envirocare of Utah facility is available for most Class A wastes generated nationwide. The site's operator, Envirocare, applied to the state on November 1, 1999, for a license amendment to accept Class B and C waste as well. Utah regulators granted the license amendment. For the amendment to take effect, however, approvals by the state legislature and the governor are required. Envirocare has deferred seeking final state approval in part because of citizens' concerns and considerable political sensitivity to waste disposal issues (e.g., a proposed commercial spent fuel storage facility near Envirocare on the Goshute reservation).

It is notable that no new commercial disposal facilities have been opened since the Envirocare of Utah site opened in 1988. After the Low-Level Waste Policy Act made states responsible for disposal of their LLW and directed the formation of interstate compacts, the states and compacts spent about $600 million in mostly failed siting efforts (GAO, 1999, also see Sidebar 2.1). A site at Ward Valley, California, was licensed by US Ecology in 1993, but land transfer issues from the federal to state government effectively blocked that site's startup. Recently, however, the Texas legislature and governor have approved bills to allow commercial LLW disposal in that state.

Although the specific reasons for the lack of success vary among compacts and states, there are several common threads. One thread is the controversial nature of nuclear waste disposal, which often manifests itself in the form of skepticism about and opposition to disposal facilities by members of the public and political leaders. Waste generators, compacts, and states have in recent years reassessed their need for disposal facilities and deferred the development of facilities because of the declining volume of Class B and C wastes, the high cost of developing new disposal

SIDEBAR 3.4
Regulation and Disposal of LLW in Near-Surface Facilities

The USNRC and the states govern the siting, operation, and closure of all LLW disposal facilities. The USNRC has set forth requirements to protect people from releases from the site, prevent inadvertent intrusion into the waste, protect workers during operation, and ensure the stability of the site after closure.

USNRC regulations for required low-level waste disposal time periods. The USNRC requires that Class A LLW be contained for up to 100 years, Class B waste for 300 years, and Class C waste for up to 500 years.

USNRC regulations for low-level waste disposal facilities. The USNRC has established technical requirements for shallow land disposal. These requirements include areas, such as wildlife preserves, to be avoided; the site must be sufficiently isolated from groundwater and surface water; and the site must not be in an area of geological activity (such as volcanoes or earthquakes). Regardless of design, all LLW disposal sites use a series of natural and engineered barriers to prevent radioactivity from reaching the environment. There are five designs for building disposal facilities: shallow land burial, modular concrete canister, below-ground vault, above-ground vault, and earth-mounded concrete bunker.

Waste treatment. Most LLW including those wastes that are LAW as defined in this report are disposed in 55-gallon drums, B-25 boxes, or other specialized concrete, metal, or sometimes wooden containers. Wastes are prepared by compaction, super compaction, dewatering solidification, consolidation, or other techniques approved by regulators of disposal sites. These requirements are spelled out in site licenses and waste acceptance plans or waste acceptance criteria.

Shallow land burial. Waste containers are placed in long, lined trenches 25 or more feet deep. The trenches are covered with a clay cap or other

facilities, and the continued availability of disposal services to most waste generators (GAO, 1999).

Current policies (specifically, surcharges and taxes levied by states that host the three commercial disposal facilities) put into place in the 1980s for managing commercial LLW have led to higher prices to generators. Potential lack of access to existing disposal capacity due to restric-

low-permeability cover, gravel drainage layers, and a topsoil layer. They then are contoured and replanted with vegetation for drainage and erosion control. In addition, an intrusion barrier, like a thick concrete slab, is added to Class C waste trenches. The sites are carefully monitored to ensure performance in compliance with the regulations. Facilities are sited in an area away from surface water and where travel of any groundwater is slow.

Other disposal systems include but are not limited to:

Modular concrete canister disposal. This method consists of individual waste containers placed within concrete canisters, which are then disposed in shallow land sites. The array of canisters has an earthen cover. This additional engineered barrier system has been used at the Barnwell, South Carolina, facility since 1995 and has been proposed for Classes B and C disposal at Envirocare.

Below-ground vault. This type of disposal uses a sealed structure built of masonry blocks, fabricated metal, concrete, or other materials that provide a barrier to prevent waste migration. It has a drainage channel, a clay top layer and a concrete roof to keep water out, a porous backfill, and a drainage pad for the concrete vault.

Above-ground vault or engineered berm. This is a reinforced-concrete building that provides isolation on the Earth's surface. Its walls and roof are two to three feet thick, and it has a sloping roof to aid water runoff. Some Canadian utilities use similar above-ground vaults for storing LLW for later disposal. For LAW, above-ground engineered berms provide the same isolation as shallow land burial. Envirocare of Utah uses above-ground engineered berms.

SOURCE: NEI <http://www.nei.org/index.asp?catnum=2&catid=73>.

tions by host states creates concerns among generators, especially in view of the planned closing of the Barnwell site to users outside the Atlantic Compact in 2008. The picture for defense LLW, much of which is radiologically similar to the civilian waste stream, is very different with access to disposal capacity being assured at a much lower cost (DOE, 2002).

Appendix A: Interim Report

4
Issues and Findings

As described in Chapters 2 and 3, low-activity wastes are regulated primarily on the basis of their origin (national defense, nuclear power, resource recovery) under a patchwork of federal and state statutes put into place over a period of almost six decades. The current system for regulating this waste lacks overall consistency and, as a consequence, waste streams having similar physical, chemical, and radiological characteristics may be regulated by different authorities and managed in disparate ways. These disparities have health, safety, and cost implications, and they may undermine public confidence in regulatory agencies.

Table 4.1 summarizes the committee's overview of the radiological hazards associated with low-activity waste and the current regulations that address the hazards. The first three waste categories shown on the table (low-level waste; slightly radioactive solid materials; and discrete radioactive sources) are governed by section 11e.(1) of the Atomic Energy Act (AEA). They meet the Nuclear Waste Policy Act's exclusionary definition of low-level waste (LLW) (see Chapter 2). In the commercial sector, waste is regulated by the Nuclear Regulatory Commission (USNRC) under 10 CFR Part 61. At Department of Energy (DOE) sites the same types of waste are controlled by DOE Order 435.1.

Radiological hazards in these first three waste categories vary greatly, however, and these differences are not adequately recognized by the broad statutory definitions of LLW. Even the USNRC's classification system for LLW (e.g., USNRC Classes A, B, and C) does not completely address these differences. At the low end, radioactivity in the very large volumes of debris, rubble, and soil is so low it is often difficult to measure.

TABLE 4.1 Summary of Low-Activity Waste Hazards and Regulations

Category	Radiological Hazard	Governing Statutes/ Regulation(s)
Low-level wastes from commercial and defense activities	Mostly short-lived (half- lives on the order of decades) fission and activation products. Some (e.g., reactor components, filters) have high specific activity and penetrating radiation. Potential short-term hazards to workers and long-term hazards to the environment if the wastes are allowed to migrate.	DOE: AEA, 11e.(1), self-regulated under DOE Order 435.1

COMMERCIAL: AEA, 11e.(1), USNRC or state regulated
—10 CFR Part 61 Classes A, B, and C per section 61.55
—Greater-than-Class C is responsibility of DOE to receive and dispose of with USNRC approval. |
| Slightly radioactive solid materials (debris, rubble, and contaminated soil from facility decommissioning and cleanup) | Mostly short-lived (half-lives on the order of decades) fission and activation products in large volumes of steel, concrete, other construction materials, and soils. Low hazards to workers but potential long-term hazards to the environment if the wastes are allowed to migrate. | |
| Discrete radioactive sources declared as waste | Mostly short-lived (half-lives on the order of decades) fission products of high specific activity. Potential short-term hazards to individuals and to the environment if the sources should make their way into metal recycle facilities or if they are allowed to migrate from waste disposal facilities. | |
| Uranium and thorium ore processing wastes | Very long-lived parent and daughter isotopes. Low specific alpha activity and little penetrating radiation. Low hazards to workers, but potential long-term hazards to the environment if the radionuclides are allowed to migrate, in particular radon gas and its daughters, which constitute an inhalation hazard. | Defense waste, pre-1978: not directly regulated

Defense waste, post-1978:
—UMTRCA, Title I
—10 CFR Part 40, Appendix A
—small quantities, under DOE Order 435.1

Commercial waste, post-1978:
—UMTRCA Title II
—10 CFR Part 40, Appendix A |

continued

TABLE 4.1 Continued

Category	Radiological Hazard	Governing Statutes/ Regulation(s)
Naturally occurring and technologically enhanced naturally occurring radioactive materials (NORM and TENORM wastes).	Very long-lived parent and daughter isotopes. Low specific alpha activity and little penetrating radiation. Low hazards to workers, but potential long-term hazards to the environment if the radionuclides are allowed to migrate, in particular radon gas and its daughters, which constitute an inhalation hazard.	DOE: DOE Order 435.1 —DOE M435.1-1, IV B.(3) covers accelerator-produced waste —DOE M435.1-1, IV B.(4) covers 11e.(2) and NORM Other: States have authority —CRCPD has recommended Part N for specific regulations.

Recognizing this, the USNRC has initiated a rulemaking on alternative dispositions for "slightly radioactive solid materials." Both the Environmental Protection Agency (EPA) and USNRC are considering allowing the use of hazardous waste landfills for these materials.[1] At the opposite extreme, discrete sources declared as waste are often extremely radioactive and have the potential to produce acute radiation effects and serious contamination incidents. The larger sources exceed USNRC Class C limits on near-surface disposal, and in the absence of a geological repository (e.g., Yucca Mountain if licensed and constructed) have no present means of disposal.

The radiological hazards in the last two waste categories in Table 4.1, uranium and thorium processing wastes and naturally occurring radioactive materials (NORM) wastes, arise from the uranium and thorium and their daughter isotopes. While their concentrations and isotopic distributions may vary, their hazards are roughly comparable. Nevertheless, their regulatory frameworks differ greatly. Uranium and thorium wastes fall under the AEA section 11e.(2) definition of byproduct materials. If the facilities that contained these wastes were under license by the USNRC at the time of the passage of the Uranium Mill Tailings Radiation Control Act (UMTRCA) in 1978, their wastes are managed according to the provi-

---

[1]Landfills for chemically hazardous wastes must meet design and permitting requirements of the EPA, under authority of the Resource Conservation and Recovery Act (RCRA). States can set standards for acceptance of radioactive materials in RCRA landfills when the state has jurisdiction.

sions of UMTRCA. Otherwise they may be managed under the Formerly Utilized Sites Remedial Action Program (FUSRAP).

Since ore residuals managed under FUSRAP were generated prior to the enactment of UMTRCA, the USNRC has determined that it does not have the authority to regulate them; such materials are not prohibited by federal law from disposal in RCRA-permitted landfills. UMTRCA wastes must be disposed in USNRC-licensed facilities. Disposal of pre-1978 ore residuals managed under FUSRAP or other programs can be regulated by the states. NORM and technologically enhanced NORM (TENORM) wastes are also regulated by the states, because they are not included in the AEA and therefore not subject to federal regulation. Among the states, NORM, TENORM, and FUSRAP wastes are not regulated consistently.

FUSRAP wastes provide a good example of political and regulatory inconsistencies. The Army Corps of Engineers is currently shipping railcar loads of FUSRAP wastes from St. Louis, Missouri, to the U.S. Ecology facility in Grandview, Idaho, which is permitted by the state for hazardous chemical wastes and radioactive materials not regulated by the USNRC. Previous FUSRAP disposals in the state-permitted Buttonwillow, California, hazardous waste landfill encountered severe opposition (see Sidebar 4.1). Another option used by the Corps is disposal at Envirocare of Utah according to that site's USNRC license for AEA 11e.(2) byproduct waste. DOE has disposed of about 1.5 million cubic meters of waste, which is mostly the same as the St. Louis FUSRAP wastes, at Weldon Springs, Missouri. This DOE facility was not an available option for the Corps.

Relative to AEA waste, NORM waste has received little attention from policy makers or the public. Sidebar 4.2 describes a situation in which NORM wastes, generally accepted for disposal at a Michigan landfill, are actually more radioactive than highly regulated LLW from the nuclear industry. In presentations to the committee, the EPA, USNRC, and the Conference of Radiation Control Program Directors clearly expressed the need for recognizing and more consistently controlling the radiological hazards of NORM wastes.

FINDINGS

In general, the committee believes that there is adequate statutory and institutional authority to ensure safe management of low-activity wastes, but the current patchwork of regulations is complex and inconsistent— which has led to instances of inefficient management practices and perhaps in some cases increased risk overall. Existing authorities have not been exercised consistently for some wastes. The system is likely to grow less efficient if the patchwork approach to regulation continues in the future.

SIDEBAR 4.1
Army Corps of Engineers FUSRAP Issues

The U.S. Army Corps of Engineers is responsible for remediating 21 sites that contain 1-2 million cubic meters mainly of uranium-contaminated soils and debris. The USNRC does not license or otherwise regulate:

- pre-1978 ore processing residuals at facilities that were not under license by the USNRC in 1978 or thereafter, or
- residuals of ores processed for other than their source material content (i.e., non-AEA section 11e.(2) material).

While the Corps believes the USNRC's legal position is correct, the position is questionable from a health, safety, and environmental perspective. Standards of individual states that control the residuals vary considerably. The above-listed residuals are radiologically and chemically similar and present similar or identical hazards to 11e.(2) byproducts, which are controlled by the USNRC. The radiological similarity between 11e.(2) byproducts and pre-1978 residuals has led some to reject the USNRC determination that the pre-1978 residuals do not come under material regulated by the USNRC and are not low-level radioactive waste.

The Corps has disposed of building rubble contaminated with pre-1978 residuals at the Buttonwillow, California, hazardous waste disposal facility. This practice was criticized in the belief that the materials should only be disposed in a USNRC licensed facility.

"When I learned that the Corps had disposed of 2,200 tons of radioactive waste in an *unlicensed* hazardous waste facility, . . . I was shocked."
Senator Barbara Boxer, Transcript, Hearing of the Senate Environment and Public Works Committee, July 25, 2000 [emphasis added— the facility was *permitted* to receive these materials, but not licensed].

Finding 1

Current statutes and regulations for low-activity radioactive wastes provide adequate authority for protection of workers and the public.

In its fact-finding meetings, site visits, and review of relevant litera-

SIDEBAR 4.2
Nuclear Power Waste Versus NORM

The Big Rock Point (BRP) nuclear power plant, located in northern Michigan, is in the midst of decommissioning. In 2001, BRP officials approached the USNRC, seeking approval for disposing of large quantities of concrete rubble from the decommissioning project in a municipal landfill in northern Michigan.

They proposed a waste characterization and monitoring protocol that would assure that no concrete rubble would go to the landfill if any appreciable quantity of radioactivity were present. All surfaces would be scanned for contamination at predetermined release limits. Any contamination would be removed. Then, the concrete would be rubbleized and bulk scanned. A 5 picocurie above background per gram of rubble cut-off value for approving or rejecting a particular load would be established. The USNRC approved the proposal under the authority of 10 CFR section 20.2002, which gives USNRC the authority to approve disposal for LLW other than in a licensed LLW facility. The plan also was approved by the Michigan Department of Environmental Quality.

The BRP personnel worked closely with the landfill owner and the township board in the rural community where the landfill is located, to assure all that the disposal of their decommissioning waste would be fully protective of the environment and the public. In general, BRP efforts were fairly successful in assuaging public concerns, though some reluctance to taking nuclear power plant waste remains in the minds of some local community residents and township board members. Michigan Department of Environmental Quality representatives had pointed out that there are other things going into the landfill that contain more radioactive material than the rubble. In fact, the coal ash that is used as daily cover for the cells show radioactive material concentrations in the range of 13 picocuries of radium per gram of ash.

Recently, the landfill operator installed portal monitors at the landfill, in preparation for accepting the decommissioning rubble. However, the portal monitor alarm has been tripped when certain loads of oil- and gas-production sludges and coal ash have been brought to the landfill. This material has been coming to the landfill for years, without any recognition of its radiological content. The landfill operator is developing operational procedures for determining when to refuse a load, which has tripped the portal alarm. The Michigan Low-Level Waste Authority has requested, and the landfill operator has agreed, to keep a log of all shipments that trip the portal alarms, to develop a better sense of radioactive materials entering the landfill.

SOURCE: Michigan Department of Environmental Quality.

ture, the committee found no instances where the legal and regulatory authority of federal and state agencies was inadequate to protect human health. This finding is consistent with that of previous studies by the National Academies and the National Council on Radiation Protection and Measurements (NCRP) described in Chapter 1 (NCRP, 2002; NRC, 1999a, 2002a). Some states, however, have chosen not to exercise regulatory authority over NORM and TENORM wastes. The USNRC has determined not to regulate certain pre-1978 uranium and thorium wastes. The EPA has so far not exercised its authority under the Toxic Substance Control Act to regulate non-AEA radioactive wastes. In addition, some wastes have not been adequately controlled in spite of the existence of regulatory authority. The EPA estimates that some 30,000 "orphan" sealed radioactive sources have disappeared from regulatory control, and notes that since 1983 there have been 26 recorded meltings of sources that were inadvertently mixed with scrap steel.[2] These incidents have been expensive, led to very conservative practices in the steel and nuclear industries, and fueled public distrust in the regulatory system (HPS, 2002; NRC, 2002a; Turner, 2003).

Finding 2

The current system of managing and regulating low-activity waste is complex. It was developed under a patchwork system that has evolved based on the origins of low-activity waste.

In its information-gathering the committee received a clear message from agencies responsible for managing and regulating low-activity waste: A more consistent, simpler, performance-based and risk-informed approach to regulation is needed (see Sidebar 4.3). Many committee members had difficulty in understanding the regulations well enough to discuss the system and its applications, as noted in Chapter 1. Similarly, the NCRP found that the current waste classification systems "are not transparent or defensible" and that the "classification systems are becoming increasingly complex as additional waste streams are incorporated into the system" (NCRP, 2002, p. 65).

Findings 3 and 4

Certain categories of low-activity waste have not received consistent regulatory oversight and management.

---

[2]The Orphan Sources Initiative is described at <http://www.epa.gov/radiation/cleanmetals/orphan.htm>.

SIDEBAR 4.3
Comments from Regulators and Managers

Radiation is radiation. Make decisions based on the radiation in the material and not based on the regulatory box of the material. **Southeast Compact Commission**

DOE would benefit from a more uniform approach to waste management, particularly when DOE uses commercial treatment and disposal. **Department of Energy**

Suggest improvements in management and oversight activities to achieve the greatest risk reductions with available resources. **Environmental Protection Agency**

Consistent, national standards for classifying radioactive materials such as pre-1978 ore processing residuals, oil and gas drilling wastes, and other NORM or TENORM, independent of pedigree. . . . **Army Corps of Engineers**

Address more consistent and harmonized regulation of like materials that fall under different regulatory regimes; identify and address opportunities for more risk-informed disposal of low-activity wastes. **Nuclear Regulatory Commission**

These comments were made by sponsors of this study at the first committee meeting.

Current regulations for low-activity waste are not based on a systematic consideration of risks.

Regulations focused on the wastes' origins have led to inconsistencies relative to their likely radiological risks. NORM and TENORM are not regulated by federal agencies because they do not fall under the AEA. State regulation of these wastes is not consistent. Nevertheless, these wastes may have significant concentrations of radioactive materials compared to some highly regulated waste streams. For example, NORM wastes routinely accepted at a landfill triggered a radiation monitor intended to ensure that rubble from a decommissioned nuclear reactor meets very strict limits on its radioactivity (see Sidebar 4.2).

Uranium mining and processing wastes, which are radiologically similar to NORM wastes, are regulated under federal authority by their status at the time UMTRCA was enacted. There are no federal regulations that prohibit ore processing residuals at facilities that were not under license by the USNRC in 1978 or thereafter from being disposed in hazardous waste facilities, but mill tailings regulated by the USNRC under UMTRCA, which may be radiologically identical to pre-1978 residuals, are prohibited from being disposed in such facilities. The disposal of FUSRAP waste in a hazardous waste facility in California has been the subject of much recent discussion in Congress, the media, and the regulatory community.

In addition to inconsistencies in regulating the radiological risks, current low-activity waste regulations generally overlook trade-offs between radiological and nonradiological risks. Hundred-thousand-cubic-yard volumes of slightly contaminated soil and debris and very heavy reactor components are being transported long distances for disposal. In developing current requirements for how low-activity wastes are managed or disposed, worker risks in excavating, loading, and unloading large-volume wastes; risks of transportation accidents; and environmental risks and costs (e.g., consuming large amounts of fossil fuel) have not been analyzed and compared in a systematic way to radiological risks.

PUBLIC CONCERNS REGARDING LOW-ACTIVITY WASTE: AN ISSUE FOR THE FINAL REPORT

On beginning this study, the committee was aware that there is persistent and widespread public concern with all aspects of radioactive waste management and disposal (NRC, 1996, 2001a, 2002a, 2003; GAO, 1999; Dunlap et al., 1993). During the committee's open sessions, members of the attending public expressed considerable lack of trust in the low-activity waste regulatory system due to its complexity, inflexibility, and inconsistency. These factors have apparently raised doubts about the system's capability for protecting public health. The key concerns raised in the open sessions—distrust of regulatory institutions and processes, the complexity of the problem, apprehension about risks, and the desire for greater stakeholder and public involvement—is consistent with a large and growing literature on public views of radioactive wastes and how to manage them (DOE, 1993; Dunlap et al., 1993; Slovic, 1993; Rosa and Clarke, 1999; Cvetkovich et al., 2002; Mohanty and Sagar, 2002; NRC, 2003).

The task of this interim report was to develop an overview of current regulatory and management practices for low-activity waste, and thus set the stage for the committee's final report, which will assess policy and technical options for improving the current practices. The assessments will include risk-informed options, and the committee strongly believes that issues of public trust and risk perception will be important considerations in the final report.

Appendix A: Interim Report

References

CID (Central Internet Database). 2003. A database that describes Department of Energy waste and site cleanup programs. Available at: <http://cid.em.doe.gov>.

Cvetkovich, George, Michael Siegrist, Rachel Murray, and Sarah Tragesser. 2002. New Information and Social Trust: Asymmetry and Perseverance of Attributions about Hazard Managers. Risk Analysis 22:359-367.

DOE (Department of Energy). 1993. Earning Public Trust and Confidence: Requisites for Managing Radioactive Wastes. Final Report of the Secretary of Energy Advisory Board Task Force on Radioactive Waste Management. Washington, D.C.: Office of Environmental Management. November.

DOE. 1996. Closing the Circle on the Splitting of the Atom: The Environmental Legacy of Nuclear Weapons Production in the United States and What the Department of Energy Is Doing About It. DOE/EM-0266. Washington, D.C.: Office of Environmental Management.

DOE. 1997. Integrated Data Base Report—1996: U.S. Spent Nuclear Fuel and Radioactive Waste Inventories, Projections, and Characteristics. DOE/RW-0006, Rev. 13. Washington, D.C.: Office of Environmental Management. December.

DOE. 2001. Summary Data on the Radioactive Waste, Spent Nuclear Fuel, and Contaminated Media Managed by the U.S. Department of Energy. Washington, D.C.: Office of Environmental Management.

DOE. 2002. The Cost of Waste Disposal: Life Cycle Cost Analysis of Disposal of Department of Energy Low-Level Radioactive Waste at Federal and Commercial Facilities. Report to Congress. Washington, D.C.: Office of Environmental Management. July.

DOE. 2003. United States of America National Report: Joint Convention on the Safety of Spent Fuel Management and on the Safety of Radioactive Waste Management. DOE/EM-0654. Washington, D.C.: Office of Environmental Management. May.

Dunlap, Riley E., Michael E. Kraft, and Eugene A. Rosa (eds.). 1993. Public Reactions to Nuclear Waste: Citizens' Views of Repository Siting. Durham, NC: Duke University Press.

GAO (General Accounting Office). 1999. Low-Level Radioactive Wastes: States Are Not Developing Disposal Facilities. GAO/RCED-99-238. Washington, D.C.: GAO.

GAO. 2000. Low-Level Radioactive Wastes: Department of Energy Has Opportunities to Reduce Disposal Costs. GAO/RCED-00-64. Washington, D.C.: GAO.

HPS (Health Physics Society). 2002. State and Federal Action Is Needed for Better Control of Orphan Sources. A position statement of the Health Physics Society. Available at <http://www.hps.org>.

IAEA (International Atomic Energy Agency). 2001. Code of Conduct on the Safety and Security of Radioactive Sources. IAEA/CODEOC/2001. Vienna, Austria: IAEA.

MIMS (Manifest Information Management System). 2003. A database of low-level waste shipments to commercial disposal facilities. Available at <http://mims.apps.em.doe.gov>.

Mohanty, Sitakanta, and Budhi Sagar. 2002. Importance of Transparency and Traceability in Building a Safety Case for High-Level Nuclear Waste Repositories. Risk Analysis 22:7-16.

NCRP (National Council on Radiation Protection and Measurement). 2002. Risk-Based Classification of Radioactive and Hazardous Chemical Wastes. NCRP Report No. 139. Bethesda, Maryland: NCRP.

NRC (National Research Council). 1986. Scientific Basis for Risk Assessment and Management of Uranium Mill Tailings. Washington, D.C.: National Academy Press.

NRC. 1990. Health Effects of Exposure to Low Levels of Ionizing Radiation: BEIR V. Washington, D.C.: National Academy Press.

NRC. 1996. Review of New York State Low-Level Radioactive Waste Siting Process. Washington, D.C.: National Academy Press.

NRC. 1998. A Review of Decontamination and Decommissioning Technology Development Programs at the Department of Energy. Washington, D.C.: National Academy Press.

NRC. 1999a. Evaluation of Guidelines for Exposures to Technologically Enhanced Naturally Occurring Radioactive Materials. Washington, D.C.: National Academy Press.

NRC. 1999b. The State of Development of Waste Forms for Mixed Wastes. Washington, D.C.: National Academy Press.

NRC. 2001a. Disposition of High-Level Waste and Spent Nuclear Fuel: The Continuing Societal and Technical Challenges. Washington, D.C.: National Academy Press.

NRC. 2001b. Improving the Operations and Long-Term Safety of the Waste Isolation Pilot Plant. Final Report. Washington, D.C.: National Academy Press.

NRC. 2002a. The Disposition Dilemma: Controlling the Release of Solid Materials from Nuclear Regulatory Commission-Licensed Facilities. Washington, D.C.: National Academy Press.

NRC. 2002b. Characterization of Remote-Handled Transuranic Waste for the Waste Isolation Pilot Plant. Final Report. Washington, D.C.: National Academy Press.

NRC. 2002c. Research Opportunities for Managing the Department of Energy's Transuranic and Mixed Wastes. Washington, D.C.: National Academy Press.

NRC. 2003. One Step at a Time: The Staged Development of Geologic Repositories for High-Level Radioactive Waste. Washington, D.C.: The National Academies Press.

Rosa, Eugene A. and Donald L. Clark, Jr. 1999. Historical Roots to Technological Gridlock: Nuclear Technology as Prototypical Vehicle. Research in Social Problems and Public Policy 7:21-57.

Slovic, Paul. 1993. Perceived Risk, Trust, and Democracy. Risk Analysis 13:675-682.

St. Onge, R. 2003. Operating and Decommissioning LLRW: Power Generators' Perspective. Presented at the 18th Annual International Radioactive Exchange LLRW Decisionmakers' Forum and Technical Symposium. Park City, Utah. June 17-20.

Turner, R. 2003. The Metal Industry's View on Unrestricted and/or Limited Metal Recycle/Release into the Commercial Market. Presented at the 18th Annual International Radioactive Exchange LLRW Decisionmakers' Forum and Technical Symposium. Park City, Utah. June 17-20.

USNRC (U.S. Nuclear Regulatory Commission). 2002. Radioactive Waste: Production, Storage, Disposal. NUREG/BR-0216, Rev. 2. Washington, D.C.: Office of Public Affairs. May.

Appendix A: Interim Report

U.S. Nuclear Regulatory Commission (Appendix A of Interim Report)

The U.S. Nuclear Regulatory Commission (USNRC) is an independent regulatory agency established by the Congress under the Energy Reorganization Act of 1974 to ensure adequate protection of the public health and safety and the environment and to promote the common defense and security in the civilian use of nuclear materials. The USNRC scope of responsibility includes regulation of:

- Commercial nuclear power; non-power research, test, and training reactors;
- Non-Department of Energy fuel cycle facilities; medical, academic, and industrial uses of nuclear materials; and
- Transport, storage, and disposal of nuclear materials and waste.

The regulatory system established by the USNRC has its authority in legislation listed in Chapter 2, Table 2.1. To fulfill this agency's Congressionally mandated mission, the USNRC has established licensing procedures for regulating the use of byproduct, source, and special nuclear materials. Specifically, the goals for radioactive waste management are to: ensure treatment, storage, and disposal of waste produced by civilian use of nuclear materials in ways that do not adversely affect future generations; and to protect the environment in connection with civilian use of source, byproduct, or special nuclear materials through the implementation of the Atomic Energy Act and the National Environmental Policy Act.

Current Nuclear Regulatory Commission (NRC 10 CFR Part 20)

Regulations define Source Materials, Byproduct Materials, and Special Nuclear Materials as follows:

Source material means:

(1) Uranium or thorium or any combination of uranium and thorium in any physical or chemical form; or

(2) Ores that contain, by weight, one-twentieth of 1 percent (0.05 percent), or more, of uranium, thorium, or any combination of uranium and thorium. Source material does not include special nuclear material.

Byproduct material means:

(1) Any radioactive material (except special nuclear material) yielded in, or made radioactive by, exposure to the radiation incident to the process of producing or utilizing special nuclear material; and

(2) The tailings or wastes produced by the extraction or concentration of uranium or thorium from ore processed primarily for its source material content, including discrete surface wastes resulting from uranium solution extraction processes. Underground ore bodies depleted by these solution extraction operations do not constitute "byproduct material" within this definition.

Special nuclear material means:

(1) Plutonium, uranium enriched in the isotope 233 or in the isotope 235, and any other material that the Commission, pursuant to the provisions of section 51 of the Act, determines to be special nuclear material, but does not include source material; or

(2) Any material artificially enriched by any of the foregoing but does not include source material (10 CFR 20.1003).

The USNRC conducts licensing and inspection activities associated with domestic nuclear fuel cycle facilities, uses of nuclear materials, transport of nuclear materials, management and disposal of low-level waste (LLW) and high-level waste (HLW), and decontamination and decommissioning of facilities and sites. USNRC also is responsible for establishing the technical basis for regulations, and provides information and technical basis for developing acceptance criteria for licensing reviews.

An important aspect of the USNRC regulatory program is its inspection and enforcement activities. The USNRC has four regional offices (Region I in King of Prussia, Pennsylvania; Region II in Atlanta, Georgia; Region III in Lisle, Illinois; and Region IV in Arlington, Texas), that conduct inspections of licensed facilities including nuclear waste facilities. USNRC also has an Office of State and Tribal Programs, which establishes and maintains communication with state and local governments and Tribes, and administers the Agreement States Program.

An Agreement State is a state that has signed an agreement with the USNRC allowing the state to regulate the use of radioactive material within that state, consistent with the USNRC regulations. Out of the 50 states, 33 are Agreement States.

USNRC issues guidance on how to implement its regulations in the form of Regulatory Guides and Staff Positions. The USNRC staff develops Regulatory Guides to establish a standard approach to licensing. They are not intended to be regulatory requirements, but they do reflect methods, procedures, or actions that would be considered acceptable by the staff for implementing specific parts of USNRC regulations.

Regulatory Guides describe the standard format and content for license applications. Staff Positions are divided into two general types: so-called "generic" positions, dealing with issues which relate to licensing activities for nuclear facilities independent of the technology or site selected; and site-specific positions, which give site guidance or advice applicable to a specific site.

In addition to the guidance, the USNRC staff uses Standard Review Plans (typically, a "NUREG" document), which provide guidance to the USNRC staff in reviewing licensee submittals. These plans are made public so that licensees and applicants understand what is needed to comply with regulations. In this respect, the licensees and applicants have this third type of guidance to assist them in preparing their demonstration of compliance with the applicable regulations and standards.

Important guidance for radiation protection programs is provided in International Commission on Radiation Protection (ICRP) and the National Council on Radiation Protection and Measurements (NCRP) technical guidelines. Applicable recommendations are cited in USNRC staff documents, which focus on dose assessments.

USNRC regulations that affect the management of low-activity waste include the Low-Level Waste Disposal Regulations (10 CFR Part 61), Radiation Protection Standards (10 CFR Part 20), and criteria related to the disposition of uranium mill tailings (10 CFR Part 40, Appendix A). The USNRC regulates the radioactive characteristics of LLW materials acceptable for near-surface land disposal through a combination of prescriptive and performance-based requirements. Performance assessment is required to calculate worker and public exposure risks associated with waste disposal. According to the USNRC, a near-surface disposal facility is one in which radioactive waste is disposed within the upper 30 meters of the land surface. Institutional control of access is required for 100 years,

TABLE A.1 Near-Surface Disposal for Allowable Concentrations of
Long-Lived Radionuclides

Radionuclide	Concentration, curies per cubic meter (Ci/m^3)
C-14	8
C-14 in activated metal	80
Ni-59 in activated metal	220
Nb-94 in activated metal	0.2
Tc-99	3
I-129	0.08
	Concentration, nanocuries per gram (nCi/g)
Alpha emitting transuranic nuclides with half-life greater than 5 years	100
Pu-241	3,500
Cm-242	20,000

SOURCE: Code of Federal Regulations, Title 10, Section 61.55.

and within 500 years radioactivity must decay to a sufficiently low level
so that it will not pose unacceptable hazards to an intruder or the general
public.

To meet this latter requirement, further prescriptive regulations define three classes of waste that are deemed suitable for near-surface disposal. Classification as Class A (the easiest to dispose), Class B, or Class C
depends on which radionuclides are present and their concentrations (see
Tables A.1 and A.2). If the waste qualifies as transuranic or is contaminated above certain limits with long-lived radionuclides, it is not suitable
for near-surface disposal.[1]

---

[1]Mining industry waste is excluded from this requirement.

TABLE A.2 Allowable Concentrations of Short-Lived Radionuclides for Near-Surface Disposal

Radionuclide	Class A Waste (Ci/m^3)	Class B Waste (Ci/m^3)	Class C Waste (Ci/m^3)
Total of all nuclides with less than 5-year half-life	700	[a]	[a]
H-3	40	[a]	[a]
Co-60	700	[a]	[a]
Ni-63	3.5	70	700
Ni-63 in activated metal	35	700	700
Sr-90	0.04	150	7,000
Cs-137	1	44	4,600

[a]There are no limits for these radionuclides in Class B or C wastes. Practical considerations such as the effects of external radiation and internal heat generation on transportation, handling, and disposal limit the concentrations for these wastes.

NOTE: Not all Class C-or-less wastes will be acceptable at all sites and some greater than Class C wastes may be acceptable at certain sites. This distinction is the essence of the difference between waste classification and site-specific decisions on remediation.

SOURCE: Code of Federal Regulations, Title 10, Section 61.55.

Appendix A: Interim Report

The Environmental Protection Agency
(Appendix B of Interim Report)

More than a dozen major statutes or laws form the legal basis for the programs of the Environmental Protection Agency (EPA). EPA authority to develop radiation protection standards and to regulate radioactive materials, including TENORM, is derived from a number of those federal laws, plus Executive Orders.

The authority to develop Federal guidance for radiation protection was originally given to the Federal Radiation Council (FRC) by Executive Order 10381 in 1959 as an offshoot of authorities of the Atomic Energy Act (42 U.S.C. 2011 et seq.) (1954). Over the next decade the FRC developed Federal guidance ranging from guidance for exposure of the general public to estimates of fallout from nuclear weapons testing. Federal guidance developed by the FRC provided the basis for most regulation of radiation exposure by Federal and state agencies prior to the establishment of the EPA.

In 1970, the responsibility for developing federal guidance for radiation protection was transferred from the FRC to the newly formed EPA under Reorganization Plan No. 3. Federal Guidance Documents are signed by the President and issued by EPA. By signing these, the President provides a framework for federal and state agencies to develop regulations that ensure the public is protected from the harmful effects of ionizing radiation. Federal Guidance is also an opportunity for the President to promote national consistency in radiation protection regulations. For example, the guidance document "Radiation Protection Guidance to Federal Agencies for Occupational Exposure," issued by EPA in 52 CFR Part 2822, January 27, 1987, established general principles and specifies the numerical primary guides for limiting worker exposure to radiation. EPA, working in coordination with agencies of the governmental Interagency Steering Committee on Radiation Standards (ISCORS), has been revising its "Federal Radiation Protection Guidance for Exposure of the General Public" for issuance in the near future; that document last published in 1960, was revised in draft in 1994, and has been undergoing significant revisions since that time.

EPA regulates radon and radioisotope emissions through its authority under the Clean Air Act (42 USC 7401 et seq.) (1970). Regulations promulgated by the Agency that control radioactive facilities and sites include 40 CFR Part 61:

- Subpart B, Underground Uranium Mines
- Subpart H, Department of Energy Facilities
- Subpart I, Certain non-DOE Facilities
- Subpart K, Elemental Phosphorous Plants
- Subpart Q, DOE Facilities Radon Emissions
- Subpart R, Radon from Phosphogypsum Stacks

Under the Radon Gas and Indoor Air Quality Research Act (USC 42 et seq.) (1986) and Indoor Radon Abatement Act (1988), as well as authorities of the Clean Air Act, EPA has developed guidance for control of radon in buildings and schools. The guidance for radon has been generally adopted as a standard for use in establishing cleanups of radioactively contaminated sites. Although indoor radon exposures are believed by the radiation protection community to be the largest radiation related risk, indoor radiation does not arise from the low-activity wastes dealt with in this report.

The Clean Water Act's (CWA) (33 USC 121 et seq.) (1977) primary objective is to restore and maintain the integrity of the nation's waters. This objective translates into two fundamental national goals: eliminate the discharge of pollutants into the nation's waters, and achieve water quality levels that are fishable and swimmable. Under this law, EPA is given the authority to establish water quality standards and regulate the discharge of pollutants into waters of the United States. Section 502(6) of the CWA includes radioactive materials in the definition of pollutants. EPA's implementing regulations at 40 CFR 122.2, which define the term pollutant, include radioactive materials except those regulated under the Atomic Energy Act. Thus EPA currently regulates radionuclides and radiation in discharges and establishes water quality standards. This includes TENORM radionuclides with the exception of uranium and thorium.

The Safe Drinking Water Act (SDWA) (42 USC 300f et seq.) (1974), is the main federal law that ensures the quality of Americans' drinking water. Under SDWA, EPA sets standards for drinking water quality and oversees the states, localities, and water suppliers who implement those standards. Implementing regulations for 40 CFR Part 141 include the establishment of national primary drinking water standards which currently include maximum contaminant limit goals (MCLG) and maximum contaminant limits (MCL) for radiation and radionuclides; current standards include radium-226 and radium-228, uranium, combined alpha, and beta and photon emitters. MCLs have also been proposed for Radon.

The Comprehensive Environmental Response, Compensation and Liability Act (CERCLA) (42 USC 9601 et seq.) (1980) and the Superfund Amendments and Reathorization Act (SARA) (42 USC 9601 et seq.) (1986) created a tax on the chemical and petroleum industries and provided broad Federal authority to respond directly to releases or threatened releases of hazardous substances that may endanger public health or the environment. CERCLA established prohibitions and requirements concerning closed and abandoned hazardous waste sites; provided for liability of persons responsible for releases of hazardous waste at these sites; and established a trust fund to provide for cleanup when no responsible party could be identified. EPA has determined that radiation is a carcinogen and thus a hazardous substance. Under the National Oil and Hazardous Substances Contingency Plan, EPA has issued guidance on removals and clean up of radioactively contaminated sites. Implementing regulations for the NCP are found at 40 CFR Part 300.

The Toxic Substances Control Act (TSCA) (15 USC 2601 et seq.) (1976) was enacted by Congress to give EPA the ability to track the 75,000 industrial chemicals currently produced or imported into the United States. EPA repeatedly screens these chemicals and can require reporting or testing of those that may pose an environmental or human-health hazard. EPA can ban the manufacture and import of those chemicals that pose an unreasonable risk. While radionuclides are considered toxic substances under the act, source material, special nuclear material, or byproduct material (as such terms are defined in the Atomic Energy Act of 1954 (42 USC. 2011 et seq.) and regulations issued under such Act) are excluded from coverage. Consequently, TENORM radionuclides may be subject to this law.

The Resource Conservation and Recovery Act (RCRA) (42 USC 321 et seq.) (1976) gave EPA the authority to control hazardous waste. This includes the generation, transportation, treatment, storage, and disposal of hazardous waste. RCRA also set forth a framework for the management of nonhazardous solid waste. The 1986 amendments to RCRA enabled EPA to address environmental problems that could result from underground tanks storing petroleum and other hazardous substances. RCRA focuses only on active and future facilities and does not address abandoned or historical sites (see CERCLA). The Hazardous and Solid

Waste Amendments (HSWA) are the 1984 amendments to RCRA that restricted land disposal of hazardous waste. Some of the other mandates of this strict law include increased enforcement authority for EPA, more stringent hazardous waste management standards, and a comprehensive underground storage tank program. RCRA specifically excludes source, special nuclear, and byproduct material from its jurisdiction. EPA's implementing regulations for RCRA do not address, but also do not prohibit, disposal of radioactively contaminated substances in landfills. With the approval of the appropriate regulatory authority, such facilities have been used for disposal of TENORM, nuclear accelerator wastes, and certain AEA materials.

Additional radiation protection authorities provided to the EPA by Congress include responsibilities for setting protective standards for radioactive waste disposal. Under the Waste Isolation Pilot Plant Land (WIPP) Withdrawal Act, as amended (P.L. 102-579, 106 Stat. 4777), Congress gave EPA the authority to regulate many of the Department of Energy's activities concerning this radioactive waste disposal site in New Mexico. EPA was required to finalize regulations which apply to all sites—except Yucca Mountain—for the disposal of spent nuclear fuel, transuranic and high-level radioactive waste. In 1998, EPA granted a certification of compliance indicating that the WIPP complied with EPA's radioactive waste disposal regulations and could open to receive these materials. The compliance criteria regulations were established by EPA in 40 CFR Part 194 and the disposal regulations set by EPA in 40 CFR Part 191.

The Energy Policy Act of 1992 (42 USC 10141 n), Section 801, required the EPA, based upon and consistent with the findings and recommendations of the National Academy of Sciences, to develop regulations on health and safety standards for protection of the public from releases from radioactive materials stored or disposed of in the proposed Yucca Mountain radioactive waste disposal site. The standards to be developed were required to prescribe the maximum annual effective dose equivalent to individual members of the public from releases to the accessible environment from radioactive materials stored or disposed of in the repository. In 1999, EPA proposed draft standards and held public hearings; final regulations were published in 2001 for use by the Nuclear Regulatory Commission and Department of Energy.

Current regulations applicable to remediation of both inactive uranium mill tailings sites, including vicinity properties, and active uranium and thorium mills have been issued by the EPA under the Uranium Mill Tailings Radiation Control Act (UMTRCA) (42 USC 2022 et seq.) of 1978, as amended. EPA's regulations in 40 CFR Part 192 apply to remediation of such properties and address emissions of radon, as well as radionuclides, metals, and other contaminants into surface and groundwater.

Appendix A: Interim Report

The McMahon Act
(Appendix C of Interim Report)

The McMahon Act (Atomic Energy Act of 1946) was focused on safeguards and security for materials that have significance in the development of "atomic fission." The Atomic Energy Act was significantly rewritten as the more familiar Atomic Energy Act of 1954. This version with several major amendments of its coverage and content comprises today's regulations from the Nuclear Regulatory Commission. Nonetheless the very first definitions that were designed to provide safeguards and security of materials involved in "atomic fission" survive with only slight changes in wording today.

The 1946 definitions were:

(b) Source Materials.

(1) Definition. The term "source materials" shall include any ore containing uranium, thorium, or beryllium, and such other materials peculiarly essential to the production of fissionable materials as may be determined by the Commission with the approval of the President.

(2) License for Transfers Required. No person may transfer possession or title to any source material after mining, extraction, or removal from its place of origin, and no person may receive any source material without a license from the Commission.

(3) Issuance of Licenses. Any person desiring to transfer or receive possession of any source material shall apply for a license therefore in accordance with such procedures as the Commission may by regulation

establish. The Commission shall establish such standards for the issuance or refusal of licenses, as it may deem necessary to assure adequate source materials for production, research or developmental activities pursuant to this Act or to prevent the use of such materials in a manner inconsistent with the national welfare.

(c) Byproduct Materials.

(1) Definition. The term "byproduct material" shall be deemed to refer to all materials (except fissionable material) yielded in the processes of producing fissionable material.

(2) Distribution. The Commission is authorized and directed to distribute, with or without charge, byproduct materials to all applicants seeking such materials for research or developmental work, medical therapy, industrial uses, or such other useful applications as may be developed, if sufficient materials to meet all such requests are not available, the Commission shall allocate such materials among applicants therefore, giving preference to the use of such materials in the conduct of research and developmental activity and medical therapy. The Commission shall refuse to distribute or allocate any byproduct materials to any applicant, or recall any materials after distribution or allocation from any applicant, who is not equipped or who fails to observe such safety standards to protect health as may be established by the Commission.

Sec. 5. (a)(1) Definition. The term "fissionable materials" shall include plutonium, uranium 235, and such other materials as the Commission may from time to time determine to be capable of releasing substantial quantities of energy through nuclear fission of the materials.

(2) Privately Owned Fissionable Materials. Any person owning any right, title, or interest in or to any fissionable material shall forthwith transfer all such right, title, or interest to the Commission.

(3) Prohibition. It shall be unlawful for any person to (a) own any fissionable material; or (b) after sixty days after the effective date of this Act and except as authorized by the Commission possess any fissionable material; or (c) export from or import into the United States any fissionable material, or directly or indirectly be a party to or in any way a beneficiary of, any contract, arrangement or other activity pertaining to the production, refining, or processing of any fissionable material outside of the United States.

(4) Distribution of Fissionable Materials. The Commission is authorized and directed to distribute fissionable materials to all applicants requesting such materials for the conduct of research or developmental activities either independently or under contract or other arrangement with the

Commission. If sufficient materials are not available to meet all such requests, and applications for licenses under section 7, the Commission shall allocate fissionable materials among all such applicants in the manner best calculated to encourage independent research and development by making adequate fissionable materials available for such purposes. The Commission shall refuse to distribute or allocate any materials to any applicant, or shall recall any materials after distribution or allocation from any applicant, who is not equipped or who fails to observe such safety standards to protect health and to minimize danger from explosion as may be established by the Commission."

Appendix B

International Approaches for Management of Low-Activity Radioactive Waste

T his appendix overviews international practices for regulating and managing low-activity radioactive wastes (LAW) as well as ongoing efforts in individual countries or internationally toward harmonizing these practices. This overview is not intended as a definitive survey of international practices, but rather to provide international perspectives for improving U.S. practices, as described in Chapter 2 and Appendix A of this report. The multiplicity of international approaches makes it difficult to develop a systematic picture—but provides fertile ground for greater exchange of ideas and information that can lead to mutual strengthening of LAW management in all countries.

The following examples have been chosen mostly from among countries that have a well-developed nuclear industry and therefore have experience with a variety of practices for managing waste. From these examples, an attempt is made to identify issues and trends relevant to strategies for LAW management and opportunities for further improvement and harmonization. This synthesis provided insights that helped the committee develop its findings and recommendations.

WASTE CLASSIFICATION

There is no internationally endorsed classification of waste at present; each country identifies its own categories of waste. This results in a diverse nomenclature (at least 20 different denominations for various waste categories exist throughout the world) that does not facilitate direct comparison. However, common features can be identified, especially for nuclear waste that falls under one of three main classification systems.

Under the first, waste is classified by its mode of disposition ("management" routes). Adopted by France, Spain, and, more recently, Japan, this classification defines four categories of waste: slightly radioactive waste (or very low level waste, VLLW), low and intermediate short-lived waste (LILW—SL), low and intermediate long-lived waste (LILW—LL), and high-level waste (HLW). These categories generally differ from one another by orders of magnitude of activity content. The distinction between short- and long-lived waste is based on the half-life (30 years) of cesium-137. However, these categories, though clearly different, are not defined a priori by generic cutoff values. These values are determined on the basis of waste acceptance criteria for a given management option when sufficient assessment results are available to allow deriving limits that are considered safe. An example of this waste classification mode is given in Table B.1.

A second classification system defines categories of waste on the basis of their main characteristics. Adopted by the United Kingdom and formerly by Germany, it more or less identifies the same broad categories of waste mentioned under the first classification, but with possible differences in the cutoff values that separate the categories. These values are defined a priori, in a generic manner; for example, the United Kingdom defines LLW as waste containing no more than 4 GBq/t in alpha emitters and 12 GBq/t in beta and gamma emitters, intermediate-level waste (ILW) as waste of low thermal output and activity of the order of 1,000 TBq/m^3,

TABLE B.1 Waste Classification in France

Activity	
Slightly radioactive	Dedicated surface disposal (Centre de stockage TFA de Morvilliers)
Low and Intermediate level	Surface disposal (Centre de Stockage de l'Aube) for wastes with half-life less than 30 years
Low and Intermediate level	Specific disposal options for wastes with half-life greater than 30 years, e.g., TENORM, graphite waste, are under study
High level	Management options under study (Law of December 30, 1991)

and HLW as waste of high thermal output (20 kW/m³) and activity concentrations ranging from 5,000 to 50,000 TBq/m³.

The third system, which considers the origin of the waste, has been adopted in some countries, among them the United States and Finland. The situation in United States is described in Appendix A. As a complementary example, the Finnish situation is interesting because, as in United States, the classification of waste according to origin leads to separate management options for waste having the same characteristics, but broad categories similar to those for the other classification approaches also are identified. In Finland, a distinction is made first between wastes from the nuclear industry, which are controlled by nuclear energy legislation, and wastes of other origins, which are controlled by radiation protection laws. These categories are both subdivided into low- and intermediate-level (LILW) waste and HLW. Disposal of LILW is at different sites according to which power supplier, Teollisuuden Voima Oy (TVO) or Imatran Voima Oy (IVO), has produced the waste. Disposal of HLW is to be in a single site in a deep geological formation.

From this short overview, one may sense the apparent complexity of worldwide classification systems for radioactive waste, but in most cases four main categories (VLLW, LILW—SL, LILW—LL, and HLW) can be identified. These approaches formed the basis for a waste classification system proposed by the International Atomic Energy Agency (IAEA, 1994), which is shown in Chapter 2 Table 2.1. The system was endorsed for publication by IAEA member states as a means to facilitate communication and exchange among countries, but it has not been incorporated directly into any country's national regulations. The system does not explicitly address naturally occurring radioactive materials (NORM) wastes (see the later section on Management of Nonnuclear Waste) and some revisions are being considered.

MANAGEMENT OF SLIGHTLY RADIOACTIVE WASTE (VLLW)

Management of VLLW generally is split into two types of practices: clearance and disposal. Clearance of waste consists of allowing waste to be freed from control, meaning that its level of activity is not of concern for radiation protection and any use can be made of the cleared waste. This practice has been adopted in many countries, for example, the United Kingdom, Sweden for metallic waste, Japan, and Spain. Germany uses two types of clearance mechanisms: free release of waste but also specific clearance, allowing higher levels of activity to be released, but restricting further disposition options to recycling or storage.

Some countries do not currently allow clearance of waste, but manage

VLLW solely by disposal in facilities approved for radioactive waste. This is the case in the United States, where surface disposal of VLLW in USNRC licensed facilities is required (although case-by-case exemptions are possible, see Chapter 2).

France does not oppose the clearance of waste for the purpose of recycling valuable material. However, this option is generally not used because of high public concern and comparatively poor economic benefit. The approach that is actually used in France begins with identifying zones in nuclear installations where products are suspected to be radioactive. All products inside these zones are thus considered radioactive; all products outside these zones are considered conventional waste and need not be subjected to further regulatory control for reasons of radiological protection. The radioactive waste content and level of activity are reconstructed through process analysis and history of operations. The validity of the estimated radionuclide content is verified by measurement. The slightly radioactive waste is then sent to a dedicated facility (the VLLW disposal facility at Morvilliers) for disposal if it meets the facility's acceptance criteria. The level of activity accepted at this site is on the order of 10 Bq/g.

The European Commission allows member states to choose whether to clear or to dispose of VLLW, but, in the case of clearance, it provides guidance on activity levels (EC, 2000a). IAEA (2004b) also has recently proposed guidelines on clearance levels, which apply to any material that contains radioactive elements. When levels are below those recommended in the guide, control of the material would not be justified for reason of radiation protection and thus can be used without restriction.

These approaches for VLLW management have been implemented and are fairly consistent among countries but essentially are used only for waste from the nuclear industry. Waste from NORM can still give rise to special considerations for its management, even at very low levels of activity, as described in the section on "Management of Nonnuclear Waste."

LILW MANAGEMENT

Disposal of LILW is widely considered to be the preferred management route and is practiced in most countries. However, the design of LILW disposal facilities may be significantly different. Further, for essentially the same types of waste, countries have chosen to implement different disposal options: surface or shallow land disposal or geological disposal (see Table B.2).

European Nordic countries and Germany have adopted deep geological disposal of LILW. There are, of course, differences in depth, design, and type of host rock among options implemented or envisaged, but all

TABLE B.2 Examples of Disposal Options for LILW

| Country | Short-lived LLW and LILW | | Long-lived LLW and LILW | |
	Option	Characteristics	Option	Characteristics
France	Surface disposal: Centre de Stockage de l'Aube	Multibarrier concepts: (i) waste package; (ii) disposal vaults, impervious cover; (iii) site features (dry disposal above water table)	Under study	Subsurface disposal of TENORM and graphite waste
Spain	Surface disposal: El Cabril[a]		Under study	
Japan	Surface disposal: Rossasho-Mura	Multibarrier concepts: (i) waste package, (ii) disposal vaults, (iii) low-permeability media, (iv) multilayer cover	Under study	
Germany	Geologic disposal: Morsleben (salt), Konrad (not licensed yet)	Accepting all waste except thermogenic ones. Disposal in cavities of large volume (salt site)		
Finland	Geological disposal: Loviisa and Olkiluoto (granite)	Multibarrier concepts: (i) Olkiluoto, two silos at 70- and 100-m depths, for LLW and ILW respectively; (ii) Loviisa, two tunnels at 110-m depth for LLW and ILW, and one cavity for decommissioning waste		
Sweden	Subsurface disposal: SFR (granite)	Multibarrier subsurface system: (i) Waste package, steel or concrete containers; (ii) disposal cavities or silo; (iii) host rock at 50-m depth below sea level		

[a]The El Cabril facility has recently begun operating specially designed cells for very low-level waste. Cell designs are based on hazardous waste regulations. Total radioactivity in the VLLW cells is restricted to be below 1 percent of the total site inventory (Zuloaga, 2003).

provide a high level of protection with regard to intrusion risks. Thus, these countries do not subdivide their LILW according to half-life. Their geological facilities generally accept all except heat-producing nuclear waste.

This is not the case for countries that have implemented surface disposal of LILW, such as the United Kingdom, Spain, France, or Japan. Sweden also uses near-surface burial at reactor sites for low-level nuclear reactor waste. The relative lack of robustness of near-surface facilities with regard to intrusion or natural events requires limiting the amount of activity that may be accepted for disposal, especially of the long-lived radionuclides. Therefore, surface disposal of LILW is mainly dedicated to short-lived waste (<30 years) to allow for substantial reduction of the risk potential of the waste within the period of time during which institutional control of the facility is maintained (some 100 years).

The amount of activity accepted for surface disposal may vary from one country to another. For example, U.S. Nuclear Regulatory Commission (USNRC) Class C waste is about 10 times higher in cesium-137 and strontium-90 content than waste accepted in Centre de l'Aube (France). Such a difference does not necessarily reflect inconsistencies in disposal practices. Acceptance criteria are mostly site and design dependent and a wide variety of conditions are encountered (e.g., sites may be located in wet or desert areas, waste may or may not be located at higher levels above the water table, disposal may be in trenches or engineered vaults). Another source of difference is in the approach used to assess disposal safety and to appraise impact acceptability (the "integrated" risk approach as opposed to the separate appraisal of elements supporting acceptability). This is further addressed the sections on "Management of Nonnuclear Waste" and "Global Approaches at the International Level."

Whatever the approach, surface disposal requires that the threshold values above which waste will not be accepted in such a facility be clearly defined and that waste be managed within disposal options that are robust against events that may jeopardize waste confinement. There is quite good consistency among European countries in limiting maximum actinide concentrations to levels of a fraction of 1 Ci/t of waste. This is also consistent with the definition of U.S. transuranic waste (with actinide content above 0.1 Ci/t).

Because surface disposal of LILW is an option that has the benefits of a rather large knowledge base and wide industrial experience, reasonable agreement on the safety issues to be addressed has been obtained (EC, 1996b; IAEA, 1999).

MANAGEMENT OF NONNUCLEAR INDUSTRY WASTE: NORM, URANIUM MINE AND MILL TAILINGS, AND DISUSED SEALED SOURCES

Internationally, wastes produced by the nuclear power and defense industries generally have received careful attention in regard to keeping their risks under control in the short and long term, which has led to fairly consistent management practices among countries, but the picture is much more difficult to draw for nonnuclear radioactive wastes. The need to control NORM waste for purposes of radiological protection is a new concern. This type of waste is not associated with the nuclear industry, and, there has been little public awareness of its radiation risks. More attention, of course, has been paid to mine and mill tailings with regard to radiological protection, but historically, they generally have been regulated and controlled by different bodies than for the nuclear industry. This has led to separate considerations on how to manage risk arising from tailings versus nuclear waste.

As for spent sealed sources, their widespread distribution for various uses involves a multiplicity of stakeholders (producer, owner, user, regulator, and so on) for their management, and there is a strong dependence on the specific practices in each country. Also, the focus has been more on keeping track of the sources than on their disposal, which generally have not been considered together with management practices for nuclear waste.

However, there is definitely a growing international concern to include management of nonnuclear waste in a more consistent framework (see "Global Approaches at the International Level"). The following examples illustrate some of the current practices and areas of improvements in this field.

NORM Waste

There is obviously a distinction to be made regarding the volumes of waste to be managed: quantities that can be shipped to dedicated centers (100 to 10,000 tons) and quantities that require on-site management (extraction residues and tailings amounting to millions of tons).

In France, disposal of waste in the first category is, in most cases, in centers for industrial wastes (surface disposal in vaults). These centers are designed to accept only nonradioactive waste (i.e., waste that is of no concern for radiological protection purposes). The operator must demonstrate that NORM waste can be handled and stored safely, without requiring any specific provisions regarding radiological protection. This is accomplished by assessment of the potential occupational exposure to workers at the disposal centers. Acceptability of waste is appraised with reference to an effective dose limit of 1 mSv. There are presently no generic criteria for the amount or concentrations of radioactive isotopes that would guarantee acceptability for disposal as industrial waste.

In practice, waste is controlled at the center's entrance by external measurement of the truckload. If the alarm threshold is reached in the portal monitors, the waste must undergo detailed verifications to locate potential orphan radioactive sources. If the verifications are negative and the alarm is due to uniform distribution of radioactivity attributable to NORM, then the operator must demonstrate that the waste will not give rise to unacceptable exposures. In most cases, waste disposed in this way is of very low natural activity and in moderate quantity, so that it is unlikely to generate impacts of concern to workers or the public.

Nevertheless, ways to improve NORM waste management in France are being developed. It is planned to better identify waste producers who are likely to generate waste of concern for radiological protection and to provide guidance for the assessment of waste impact. In parallel, this may lead to defining activity levels and volumes of waste above which facilities receiving the waste should be licensed to dispose of radioactive substances. This should help to better screen waste that may require additional safety measures for its disposal. Categories of waste have already been identified (mostly coming from the rare-earth extraction industry) that are much too active to be accepted in centers for industrial waste disposal. A specific disposal option together with some nuclear waste (graphite) is being studied.

Germany's approach to NORM waste management has evolved considerably. Formerly based only on consideration of an exemption level of 500 Bq/g for the total content of naturally occurring radionuclides in material outside the nuclear industry, radiation protection issues have been more thoroughly addressed through the elaboration of a list of residues for which radiological protection may be relevant and through assessment of public exposures in the short and long term for residues of concern. From studies carried out on the subject, criteria have been derived for different options of NORM waste management. These are presented in Table B.3.

In Germany, deposits of NORM waste in very large volumes are frequently encountered for which realistic impact assessments have been performed (Goldammer, 2004). The assessments consider the long-term evolution of impacts and the possibility of intrusion onsite with the use of waste in building material. The author points out that in some cases, exposures significantly above 1 mSv cannot be ruled out for large-volume deposits, even at a concentration as low as 1 Bq/g (for each radionuclide

TABLE B.3 Criteria for TENORM Disposal in Germany

Use or Disposal Option for TENORM	Criterion in Bq/g
Use or disposal of waste rock covering an area over 1 ha in the catchment area of a usable aquifer	U-238 ≤ 0.2 and Th-232 ≤ 0.2
Disposal of more than 5.000 tons annually in the catchment area of a usable aquifer	U-238+Th-232 ≤ 0.5
Residues added to building materials with ratio above 20% (house construction) or above 50% (other construction types)	U-238+Th-232 ≤ 0.5
Other use or disposal not covered by the above cases	U-238+Th-232 ≤ 1
Underground disposal	U-238+Th-232 ≤ 5

of uranium and thorium decay chains), and questions such concentration being recommended internationally as a general exclusion level. Indeed, Germany recommends threshold values of 0.2 Bq/g for large deposits of waste rock. That value is also consistent with the general exemption levels adopted by South Africa, which faces problems of management of huge quantities of residues from mineral extraction, and has set a limit of 0.2 Bq/g for radium-bearing deposits, below which regulation for radiological protection purposes can be disregarded.

Uranium Mine and Mill Tailings

Throughout the world, millions of tons of tailings from uranium ore extraction are piled in surface areas close to where the ore was mined and processed. Most tailings show natural activity levels high enough to require that some measures be taken to ensure radiological protection of the public and the environment. Mainly, two types of disposal options are encountered: Residues and waste rock are spread in layers in thalwegs or in open pits and often are contained by dams so as to protect nearby rivers from dispersion of the residues into surface waters. These may be covered by layers of comparatively low-activity material (usually waste rock of low grade) for protection against radiation, radon emissions, or airborne dust. The other broad type of disposal option is to cover the waste with a layer of water (a few meters deep), generally by flooding areas where tailings have been piled.

Each country has its own regulations with regard to radiological protection of the public from hazards arising from such disposal sites. In many cases, requirements are made for controlling exposures from the sites and monitoring them to make sure that added exposure and activity discharge into water remains acceptable.

In Europe, EC Directive 96/29 sets a limit of 1 mSv (above natural background) for public exposure, which applies to uranium tailings disposal (EC, 1996a). Demonstrating compliance with this requirement may be difficult because the discrimination between exposures from natural background radiation and those occasioned by uranium tailings disposal is not easy. However, in most cases, the disposition method (cover by rock or water) together with water control and restriction of access to the site are efficient in keeping site-related exposures comparable to background.

The main issue raised today concerns the long-term evolution of these exposures from possible on-site intrusion and loss of performance of protective covers. The issue has not yet been resolved, but there is clearly growing international concern about addressing it. IAEA (2002) has published a safety guide acknowledging these problems and clearly recommending that separate considerations be made between "historical" deposits, which may be need intervention so as to keep exposures within acceptable limits, and ongoing practices involving mining and associated waste management, which must balance occupational and public protection goals in the short and long term. The IAEA recommends, where possible, better isolation of the waste from the accessible environment, in particular in mine pits.

Whatever the case considered, the waste volume, as for some NORM waste, limits the disposition options that can be envisaged. Further, these large-volume disposition options cannot easily be compared with the options currently used for nuclear industry wastes. Nevertheless, uranium mine tailings should be part of the overall plan for achieving consistent levels of risk in managing all LAW. IAEA has recently launched work to include uranium tailings as well as nonnuclear waste disposal in a common framework, as discussed in the section on "Global Approaches at the International Level."

Disused Sealed Sources

As mentioned before, approaches to managing disused sealed sources depend on the regulatory structure in each country and a multiplicity of interests. The European Commission (2000b) report lists at least six key organizations involved in the life cycle of a source: regulator, manufacturer, original equipment manufacturer, distributor, user, and waste management organization(s). The report offers an overview of the systems of management of sealed sources in member states of the European Union (EU). Although the systems are different, there is always control by one or more of the parties involved, usually the manufacturer or the user. In most cases, the user needs to be licensed to posess sealed sources, but the level of control from regulators, once a license has been approved, is disparate from one member state to another. The number of sources lost does not seem to reflect the regulatory structure in EU member states, but the authors of the report point out areas of improvement and good practices in management systems.

One of the main problems raised for the management of disused sealed sources is the risk of bankruptcy of the user (licensee), thus breaking the management chain. Preventive measures can include a fund for managing orphan sources, as implemented in France, or an annual license fee discouraging users from holding sources for a long period without considering their disposition route, as applied in Finland. The report also points out that the risk of loss of disused sources left for storage at users' premises is increasing and that there is a need to focus on the management of sources of higher hazard potential.

Areas identified for improvement include harmonizing practices among countries and avoiding dilution of responsibilities, hence, more centralized practices and control are needed. In particular, the report outlines the benefits of creating national databases that allow separate recording of sources in- and out-of-service; implementing centralized interim storage facilities; and issuing a common code of practices among countries since transboundary movement of sources is frequent. It is also believed that system efficiency would be enhanced when under the control of one organization or a lead regulator.

Very few countries have developed definitive disposal options for sealed sources. Further, there are many instances in which lost sources have caused serious injuries, or where safety conditions for storage or disposal of sources are poor. To help remedy this, international practices encourage the return of sources to their manufacturers from users in countries where elimination routes are not likely to become available, e.g., developing countries. IAEA advises countries in the safe management of sources and has implemented case-by-case storage solutions where sources cannot be removed. There is need to establish guidance and common practices for the international shipment of these sources as well as their storage and disposal.

GLOBAL APPROACHES AT THE INTERNATIONAL LEVEL

Important efforts have been made by the international community to achieve a unified system for protecting workers and the public from the hazards of ionizing radiation. The Basic Safety Standards (BSS) (IAEA, 1996) is a worldwide reference. The BSS sets out the concepts of exclusion of exposures that are not amenable to control, exemption of practices that are not relevant to radiation protection dispositions, and clearance of radioactive material for which regulatory control can be relaxed. These concepts are fundamental for regulating and managing LAW.

The BSS makes a clear distinction between requirements relevant to normal practices and those relevant to intervention situations (where actions must be taken to reduce exposures resulting from accidents or from some past practices). For this latter case, levels are proposed in the BSS for which the implementation of intervention actions is recommended. As for practices, the BSS recommends that they be justified and that effective doses incurred from normal activities involving radioactive substances do not exceed 20 mSv/year (averaged over five years and not exceeding 50 mSv in a single year) for the worker, and 1 mSv for the relevant critical groups of the public. BSS also requires optimization of practices with regard to radiation protection, in the sense that individual doses must be kept as low as reasonably achievable (ALARA), economic and social factors being taken into account.

European Commission Directive 96/29, which supersedes national regulation in EU member states (25 countries concerned today), enforces application of the same requirements for justification of practices, optimization, and dose limitations and defines a set of exemption levels. It also requires member states to define appropriate dispositions and levels in intervention situations. The directive covers all activities involving radioactive material. It also addresses the possibility of enhanced exposure to natural radiation resulting from nonnuclear activities and requests member states to take the appropriate disposal actions to comply with the requirements set for normal practices or intervention situations.

There is thus a unified system for radiation protection that covers waste-management-related activities including the disposal of waste originating from nuclear as well as nonnuclear industries. However, there are areas that require further guidance, in particular for achieving practical criteria for the management of slightly radioactive waste, for appraising long-term radiation protection issues, and for assessing safety and bringing consistency in management of waste from all origins.

Concerning clearance levels or activity concentrations in material that may be disregarded for purposes of radiation protection, additional guidance is given by the European Commission (2000a) Report 122 and IAEA Safety Guide RS-G-1.7 (IAEA, 2004b), as explained in the section on "Management of Slightly Radioactive Waste (VLLW)."

Concerning the application of radiation protection requirements to potential long-term exposures that are specific to radioactive waste disposal, the most definitive guidance in this field is given by the International Commission on Radiological Protection (ICRP, 1998) in its Publication 81

(ICRP-81). As a broad summary, ICRP recommends applying "constrained optimization," rather than dose limitation, for achieving protection of the public from the long-term hazards occasioned by waste disposal. These recommendations clearly acknowledge the difficulty that is specific to long-term evolution of waste disposal. Shifting from a dose limitation system to a constrained optimization system is relevant to the fact that, for disposal, only projections of future doses to the public can be made. Dose limitation implies control of real exposures from a particular practice and possible action to reduce them, in particular for optimization purposes. However, one cannot rely on this control for long-term disposal. Thus, optimization must be made a priori so that doses will be ALARA.

ICRP-81 sets two conditions for optimization: The first is that disposal should be implemented through application of sound technical and managerial principles; the second is that the projected dose should be kept under, or close to, given values. For a normal situation (all barriers performing as expected, no accidental events, no intrusion, and so on), ICRP recommends applying a constraint of 0.3 mSv (consistent with the protection goal of 1 mSv, but accounting for the fact that total exposures to the public may come from disposal as well as other sources).

ICRP also gives guidance for appraising the "acceptability" of exposures potentially incurred in case of intrusion into the disposal area and recommends consideration of two values of effective dose: 10 mSv and 100 mSv. ICRP considers these values to be indicators for appraising the level of safety achieved by disposal in case of future intrusion, since intruders are considered to have no knowledge of the site and thus do not deliberately intrude. In this sense, future intrusion may be considered similar to a situation occurring today, where people may be subjected to exposures from unknown disposal sites on which they have accidently intruded, and possibly calling for intervention. According to ICRP, intervention is rarely needed when exposures are below 10 mSv, whereas it is almost always required when they exceed 100 mSv.

Finally, although guidance can be found in many fields of interest for the management of radioactive waste, there are still discrepancies in the application of safety measures and assessments, depending on the category of waste considered, which cause difficulties in appraising whether the disposal routes implemented or envisaged for each category are appropriate. Important guidance can be found concerning methods and requirements for safety assessments in IAEA safety standards. A joint convention on the safety of spent fuel management and the safety of radioactive waste management was adopted in 1997, obliging each contracting party to apply common safety standards and to report on the consistent implementation of waste management (IAEA, 1997). However, these are not sufficient to establish a strategy for radioactive waste management that would clearly define all elements necessary to achieve consistency in this field. That is why IAEA has recently launched the development of a common framework for the disposal of radioactive waste, aimed at providing a basis on which radioactive waste can be classified, identifying appropriate generic waste disposal options for each category, and defining the means for achieving safety of disposal options.

Similar strategies for waste management can be found at the national level. For example, France's national plan for radioactive waste management encompasses all types of waste (except HLW, which is addressed within the framework of specific legislation), regardless of origin, and aims at achieving consistency in its approaches. The mandate for the French Safety Authority for Nuclear Facilities is to involve political representatives, associations, institutional stakeholders (other regulators, expert organization, national agencies), and waste generators in the planning process.

SUMMARY

There is rather good consistency in the management options adopted internationally for LLW coming from the nuclear industry. Operational solutions for VLLW are now available (disposal or clearance) and surface disposal options are generally consistent in the sense that the categories of waste liable to be accepted at surface centers are fairly comparable among countries that implement this management route. There might be substantial differences in the preferred design options, but in all cases, LLW management includes thorough safety assessments to account for long-term risks.

There remain inconsistencies in the activity levels of long-lived waste that can be accepted for surface disposal. Since surface disposal is not robust in the long term against intrusion and natural risks, these levels should be rather homogeneous, which is not necessarily the case today. Differences of several orders of magnitude may exist in long-lived radionuclide concentrations in VLLW, LLW, and nonnuclear waste allowed for surface disposal in various countries. There is therefore a need for greater harmonization. If discrepancies exist, they should be justified by demonstrating that the options chosen are optimal. For instance, limits could be justified so as to accommodate waste with long-lived content that cannot be easily separated from short-lived waste, disposal of some waste in available facilities may be preferred so as to avoid safety problems of interim storage, and so on.

NORM waste and uranium mill tailings management are of concern from the viewpoint of radiological protection. Even if solutions for the near-term protection of the public and the environment are adequate, there is a need to better consider long-term risks associated with disposal—consistent with the requirements for nuclear waste. However, the very large volumes involved may require specific considerations so as to define optimal solutions for their safe management. Among other issues, this should lead to defining levels for NORM waste that may be disregarded within respect to radiological protection needs. On the other hand, higher levels that would require additional protection, so that such waste is unlikely to be affected by external events, also need to be identified.

Internationally, there is convergence toward considering that a value of 1 mSv of added dose to the public is an appropriate limit for normal exposures arising from waste management activities. However, it is a requirement to demonstrate that exposures are ALARA. There are significant differences in applying this principle among countries, ranging from the technological approach (designing a confinement system to be as robust as possible) to a fully integrated risk-based probabilistic approach.

The technological approach involves demonstrating the ALARA standard by designing the facility against plausible risks and showing that better technical solutions are not available without incurring undue costs. This approach is quite convincing with regard to uncertainties in meeting a risk objective, but it is based on a somewhat arbitrary expert appraisal of the design. The integrated approach has the merit of unifying a criterion of acceptability, which is helpful for discussions among stakeholders. However, it can be fragile because it is based on calculations that can be revised over time (probabilities of long-term evolution are very difficult to assess and the dose limits or parameters to calculate it may vary).

Whatever approach is preferred, demonstrating ALARA always involves nontechnical arguments. Hence, a key step is to involve the public and stakeholders in ALARA decisions. In arriving as such decisions, it seems important to distinguish ongoing practices from remedial actions (intervention) considering that actions to avoid hypothetical doses in the future may result in unnecessary exposures to workers and other societal impacts. A good example would be applying ALARA to NORM waste and uranium mill tailings, where interventions involving of millions of tons of waste obviously would be difficult.

There is clearly growing interest in harmonizing management of waste from all sources and achieving a consistent framework in which generic waste management solutions can be identified to establish a consistent policy for disposing of all types of waste and providing adequate answers for industrial needs. All stakeholders' involvement will be required to achieve such a framework.

Appendix C

Presentations to the Committee

Washington, D.C., December 4-5, 2002

The Nuclear Regulatory Commission's regulation of low-activity wastes and expectations for this study, Scott Flanders, USNRC

The Department of Energy's regulation of low-activity wastes and expectations for this study, Karen Guevara, DOE

The Southeast Compact Commission's role in managing low-activity wastes and expectations for this study, Mike Mobley, SECC

The Army Corps of Engineers' role in managing low-activity wastes and expectations for this study, Tomiann McDaniel and John MacEvoy, USACE

The Environmental Protection Agency's regulation of low-activity wastes and expectations for this study, Adam Klinger, EPA

Public comments

Richland, Washington, February 6-7, 2003

Introduction and overview of the DOE Hanford's low-level waste burial grounds, Rudy Guercia, DOE-Richland

Hanford Environmental Restoration Disposal Facility (ERDF), Owen Robertson, DOE-Richland

Views of the Hanford Advisory Board, Ken Bracken, HAB

Roundtable discussion led by David Leroy, Committee Chairman

Public comments

Hanford Site Visit

U.S. Ecology briefing and site tour, Mike Ault, U.S. Ecology

ERDF briefing and site tour, Rudy Guercia, DOE-RL

200 West Area low-level waste burial site tour, Rudy Guercia, DOE-RL

Salt Lake City, Utah, April 16-17, 2003

Comments from the Tooele County Commissioners, Gene White, Commissioner

Comments from the Utah Department of Environmental Quality, Bill Sinclair, Division of Radiation Control

International Uranium Corporation: Overview and waste issues, Dave Frydenlund, IUC

National Mining Association perspective, Tony Thompson, NMA (by telephone)

Public comments

Envirocare of Utah Site Visit

Overview and discussion, Ken Alkema, Envirocare of Utah

Bus tour of the site, Gene Perry, Envirocare of Utah

Washington, D.C., June 11-13, 2003

Risk-based classification of radioactive and hazardous chemical wastes—NCRP 139, Allen Croff, Oak Ridge National Laboratory

Perspectives from the Conference of Radiation Control Program Directors on medical waste and NORM, Jill Lipoti, New Jersey Department of Environmental Protection

Increasing disposal options for low-activity and mixed wastes, Adam Klinger, EPA

Disposition of slightly radioactive solid materials, Frank Cardile, USNRC

Milestones and millstones: Industry experience with low-activity waste disposals, Paul Genoa, Nuclear Energy Institute. Comments by Alan Pasternak, CalRad Forum (by telephone)

Roundtable discussion: Framing recommendations for changes in regulatory policy, Frank Marcinowski, EPA; Lawrence Kokajko, USNRC; Karen Guevara, DOE; Kathryn Haynes, SECC

Perspectives on low-activity waste issues, Diane D'Arrigo, Nuclear Information and Resource Service; Judith Johnsrud, Sierra Club

Public comments

Paris, France, September 22-25, 2003

National Plan for Radioactive waste management in France, Jérémie Averous, Direction Générale de la Sûreté Nucléaire et de la Radioprotection (DGSNR)

LLW disposal management and safety in France, Arnaud Grevoz, Agence Nationale pour la Gestion des Déchets Radioactifs (ANDRA)

Mine tailing management and impact in France, Anne Christine Servant, Institut de Radioprotection et de Sûreté Nucléaire (IRSN)

Radiological protection principles evolution: Application to waste management (Jean François Lecomte, IRSN and International Commission for Radiation Protection (ICRP)

Waste management regulation in EU, Derek Taylor, European Commission

Site visit to the LLW disposal facility Centre de l'Aube

Site visit to the LAW disposal facility at Morvilliers

LLW management in Japan, Atsu Suzuki, Nuclear Safety Commission (NSC)

LLW management in South-Korea, Sang Hoon Park, Korean Institute of Nuclear Safety (KINS)

LLW management in Spain, Pablo Zuloaga, Empresa Nacional de Residuos Radioactivos S.A (ENRESA)

LLW management in Belgium, Jean-Paul Minon, Organisme National des Déchets Radioactifs et des matières Fissiles enrichies (ONDRAF)

LLW management in Germany, Bruno Baltes, Gesellschaft für Anlagen und Reaktorsicherheit (GRS)

A common framework for radioactive waste disposal, Philip Metcalf, International Atomic Energy Agency (IAEA)

Washington, D.C., November, 29, 2004

Recap of Senate Energy and Natural Resources Hearing on Low-Level Wastes (LLW), Pete Lyons, Clint Williamson (Domenici); Jonathan Epstein, Sam Fowler (Bingaman)

Sponsors' perspectives and suggestions for completing the study:

Environmental Protection Agency, Adam Klinger, Dan Schultheisz

Army Corps of Engineers, Tomiann McDaniel

DOD Executive Agent for Low-Level Radioactive Waste, Richard Conley (via telephone)

The Institute for Applied Energy—Japan, Shigenobu Hirusawa

California Environmental Protection Agency, Jeffrey Wong (via videoconference)

Southeast Compact Commission, Mike Mobley

Midwest Interstate Low-Level Radioactive Waste Compact, Ron Kucera (via telephone)

Department of Energy, David Mathes

Nuclear Regulatory Commission, Scott Flanders

Public Comments

Appendix D

Committee Biographies

CHAIRMAN

David H. Leroy has his own law practice in Boise, Idaho, which specializes in governmental and administrative law issues. He has extensive experience in the legal, policy, and political arenas. As an appointee of President George H. Bush, he was confirmed by the Senate in August 1990 as the first U.S. waste negotiator, a post created by Congress in the 1987 Waste Policy Amendments Act to assist the government in siting a geologic repository for high-level waste. In 1993 Mr. Leroy turned his attention to low-level waste, especially the general failure of the 1980 Low-Level Waste Policy Act. Recently he has sought to develop improved technical and public policy solutions for managing low-level waste, including the assured storage concept. Before his appointment as waste negotiator, he served as Lieutenant Governor of Idaho and Idaho Attorney General. He has made numerous presentations and authored a variety of publications, including reports on low-level waste disposal, repository siting, and negotiation. Mr. Leroy received his B.S. in 1969 and J.D. in 1971 from the University of Idaho, and Master of Laws in Trial Practice and Procedure in 1972 from New York University School of Law.

VICE CHAIRMAN

Michael T. Ryan is an independent consultant in radiological sciences and health physics. He is an adjunct associate professor in the College of Health Professions at the Medical University of South Carolina. He is also an adjunct faculty member at the Charleston Southern University and the College of Charleston. Dr. Ryan is editor-in-chief of *Health Physics Journal*. Recently he was appointed by the Nuclear Regulatory Commission to a four-year term (2002-2006) as a member of the Advisory Committee on Nuclear Waste. In addition, he is currently serving on the Scientific Review Group appointed by the Assistant Secretary of Energy to review the ongoing research in health effects at the former Soviet weapons complex sites the Southern Urals and on two committees of the National Academies. In 1996-1997 Dr. Ryan was the vice president of Barnwell Operations for Chem-Nuclear Systems, Inc., where he had overall responsibility for operation of the low-level radioactive waste disposal and service facilities in Barnwell, South Carolina. From 1984 to 1996 he served as the company's director, and then vice president of regulatory affairs with the responsibility for developing and implementing regulatory compliance policies and programs to comply with state and federal regulations. Before that, Dr. Ryan spent seven years in environmental health physics at Oak Ridge National Laboratory. Dr. Ryan received his Ph.D. in 1982 from the Georgia Institute of Technology, where he was recently inducted into the Academy of Distinguished Alumni. He earned his M.S. in radiological sciences and protection from the University of Lowell, Mass. in 1976 and his B.S. in radiological health physics from Lowell Technological Institute in 1974. He is a recipient of the University of Massachusetts—Lowell's Francis Cabot Lowell Distinguished Alumni for Arts and Sciences Award.

COMMITTEE MEMBERS

Edward Albenesius retired in 1992 as manager of the advanced waste technology division and senior advisory scientist at the Savannah River Site, SC. His expertise includes treating and disposing of low-level and transuranic waste from nuclear fuel reprocessing and nuclear materials production for national defense, environmental monitoring, and health physics. He conceived and implemented the first integrated program for managing low-level wastes at a major Department of Energy (DOE) site, resulting in large reductions in waste volume and disposal in engineered facilities—departing from earlier practices of disposal in open trenches. Dr. Albenesius also held temporary assignments with the DOE where he coordinated the revision of DOE Order 5820.2A on radioactive waste management and with several task forces for the Nuclear Regulatory Commission and the National Council on Radiation Protection (NCRP). As a consultant to the International Atomic Energy Agency in 1995 he helped prepare management plans for low-activity waste and spent sealed sources for 20 developing countries. Dr. Albenesius received his Ph.D.

degree in organic chemistry from the University of North Carolina in 1952 and his A.B. degree in chemistry from the College of Charleston, SC in 1947.

Wm. Howard Arnold (NAE) retired in 1989 as general manager of the advanced energy systems division of Westinghouse Electric Company. His primary area of expertise is in the commercial nuclear fuel cycle, including nuclear power, fuel, and waste management. He has managed multidisciplinary groups of engineers and scientists working in reactor core design and led work that promoted the use of centrifuge technology in uranium enrichment. Dr. Arnold's experience includes managing residues from uranium enrichment and low-activity wastes from reactor operation and spent fuel storage. As vice president, Westinghouse Hanford Company, he was responsible for engineering, development, and project management at the Hanford Site from 1986-1989. He was elected to the National Academy of Engineering in 1974. Recently Dr. Arnold has been involved in an advisory capacity in the cleanup of DOE nuclear weapons material productions sites, especially in the vitrification plant at the Savannah River Site. Currently he is chairman of the National Academies' Committee on Improving the Scientific Basis for Managing Nuclear Materials and Spent Nuclear Fuel. He received his A.B. in 1951 from Cornell University, and his M.A. 1953 and Ph.D. in physics in 1955, both from Princeton University.

François Besnus is head of the office for safety evaluation of radioactive waste disposal in the Institute of Radiological Protection and Nuclear Safety (IRSN), Fountenay aux Roses, France. His current work includes evaluating the safety of near surface disposals of low- and intermediate-activity waste in France and participating in the development of safety standards for the European Union. Previously as a staff officer in the IRSN department for protection of man and the environment, he was in charge of very low-level and mining and milling waste management. He helped to establish French collaborations with eastern countries for assessing the extent of radioactivity migration in the Chernobyl area and for managing the large volumes of low-activity waste that resulted from the cleanup of contaminated areas. Dr. Besnus received his Ph.D. in radiochemistry in 1991, an M.S. degree in radiochemistry in 1986, and an M.S. degree in geology in 1985, all from Paris XI University.

Perry H. Charley is director of the uranium education and geographical information systems programs within the division of math, science, and technology at the Shiprock campus of Dinè College, NM, a Navajo institution. Mr. Charley has over 30 years of experience performing environmental, health impact, and psychosocial impact studies. Currently he is the principal investigator of four epidemiological research projects, the foremost being a DNA damage study of Navajo communities impacted by past uranium mining practices. From 1983 through 1999 he held several positions for the DOE and EPA uranium mill tailings remedial action (UMTRA) project, including director of the Navajo Nation's UMTRA program and the Navajo Abandoned Mine Reclamation Program. He has served on several EPA advisory committees. Mr. Charley received his B.S. degree in environmental science from the University of Arizona in 1979.

Gail Charnley is principal of HealthRisk Strategies, a consulting firm in Washington, DC. Dr. Charnley's areas of expertise are toxicology, environmental health risk assessment, and risk management science and policy. She writes and speaks extensively on issues related to the role of science and risk analysis in environmental health policy and decision-making. She is an adjunct faculty member in the Harvard School of Public Health's Center for Risk Analysis and has chaired or served on numerous peer review panels convened by the Environmental Protection Agency and the Food and Drug Administration. During its tenure, she was executive director of the Presidential/Congressional Commission on Risk Assessment and Risk Management, mandated by Congress to evaluate the role that risk assessment and risk management play in federal regulatory programs. Before her appointment to the Commission, she served as acting director of the toxicology and risk assessment program at the National Academies. She has been the project director for several National Academies committees, including the Committee on Risk Assessment Methodology and the Complex Mixtures Committee, and served as the chair of several U.S. Army Science Advisory Board committees that evaluated health risk assessment practices in the Army. Dr. Charnley received her Ph.D. in toxicology from the Massachusetts Institute of Technology in 1984 and her A.B. (with honors) in molecular biology from Wellesley College in 1977.

Sharon M. Friedman is professor of journalism and communication and director of the science and environmental writing program at Lehigh University in Bethlehem, PA. Her research and consulting activities focus on how scientific, environmental, and health risk issues are communicated to the public. Prof. Friedman chaired the DOE's Advisory Committee for its low dose radiation research program. She has served as a consultant to the President's Commission on the Accident at Three Mile Island, the United Nations Economic and Social Commission for Asia and the Pacific, and various U.S. government agencies and industries on environmental and risk communication. Elected a Fellow of the American Association for the Advancement of Science (AAAS) in 1989 for her contributions toward furthering the public understanding of science and technology, she served as a member of the AAAS Council for six years. Besides co-editing of two books, *Communicating Uncertainty: Media Coverage of New and Controversial Science* and *Scientists and Journalists: Reporting Science as News*, she has authored another book and numerous articles and book chapters. Prof. Friedman is associate editor of the journal, *Risk: Health, Safety and Environment*, and a member of the editorial advisory board of the journal, *Science Communication*. She is a member of the National Academies' Committee on Assessment of the Centers for Disease Control and Prevention's Radiation Studies. She received her M.A. in Journalism from Pennsylvania State University in 1974, a graduate certificate in public relations from American University in 1970, and her B.A. in biology from Temple University in 1964.

Maurice Fuerstenau (NAE) is professor of metallurgy at the Mackay School of Mines, University of Nevada, Reno. His expertise is in mineral extraction, processing, and hydrometallurgy. His work covers ore benefaction and dealing with residues, which include technologically enhanced naturally occurring radioactive materials. Among his numerous refereed publications and books, Dr. Fuerstenau has recently completed the volume *Principles of Mineral Processing*. He has been recognized by awards from the American Institute of Mining and Metallurgical Engineers, and by election to the National Academy of Engineering (NAE) in 1991. He has served as member, vice chair, and chair of committees of the NAE section on petroleum, mining, and geological engineering, and the NAE committee on membership. Dr. Fuerstenau received his Sc.D. in 1961 and S.M. in 1957 from the Massachusetts Institute of Technology, and his B.S. in 1955 from the South Dakota School of Mines.

James T. Hamilton is professor of public policy, economics, and political science at Duke University, where he served as associate director of the Sanford Institute for Public Policy. His expertise includes the economics of regulation, public choice in a political economy, and environmental policy. Dr. Hamilton's numerous publications include the book, *Calculating Risks: The Spatial and Political Dimensions of Hazardous Waste Policy*, co-authored with W. Kip Viscusi (MIT Press 1999). His article "Testing for Environmental Racism: Prejudice, Profits, Political Power?" *Journal of Policy Analysis and Management* 14:1 (Winter 1995) won the journal's best article of the year award. In 2001 he won the Association for Public Policy Analysis and Management's David N. Kershaw award. He earned his Ph.D. in economics in 1991 and his B.A. summa cum laude in economics and government in 1983, both from Harvard.

Ann Rappaport is a faculty member in the department of urban and environmental policy and planning at Tufts University. She held previous appointments in the department of civil and environmental engineering and in the center for environmental management at Tufts. Her work deals with both the technical and policy challenges of managing hazardous waste: health effects, site assessment and management, waste reduction and treatment, and risk assessment and management—with an emphasis on corporate responsibility and decision making. Her research has examined environmental, health, and safety programs in multinational corporations. Dr. Rappaport has published two books, several chapters, and numerous articles and reports. She was a member of the international committee of the National Advisory Council for Environmental Policy and Technology for the Environmental Protection Agency. She also served on the National Academies' Committee on Evaluation Protocols for Commercializing Innovative Remediation Technologies. Dr. Rappaport received her Ph.D. in civil engineering from Tufts University in 1992, her M.S. in civil engineering from the Massachusetts Institute of Technology in 1976, and her B.A. in Asian and environmental studies from Wellesley College in 1973.

D. Kip Solomon is an associate professor in the Department of Geology and Geophysics at the University of Utah. He specializes in fluid flow in soils and shallow aquifers, emphasizing the fate and transport of contaminants. Dr. Solomon has also worked on techniques for determining the age of shallow groundwater using tritium and helium isotopes and using these tools to examine fluid flow in porous and fractured systems. He won the outstanding faculty research award in his department in 1997-1998 and was associate editor of *Ground Water* from 1996-2001. He served on the National Academies' Panel on Conceptual Models of Flow and Transport in the Fractured Vadose Zone from 1998 to 2001. Dr. Solomon received his B.Sc. in geological engineering in 1983 and his M.Sc. in geology in 1985 from the University of Utah, and his Ph.D. in earth sciences in 1992 from the University of Waterloo.

Kimberly Thomas is deputy division leader of the chemistry division at Los Alamos National Laboratory (LANL). Her expertise includes managing wastes from research and medical isotope production. Dr. Thomas has supervised all aspects of medical isotope production at LANL. She has also directed research on accelerator transmutation of waste, geochemical behavior of radionuclides, actinide bioassay measurements, nuclear weapons debris analyses, processing of uranium ores, and fundamental actinide chemistry. She has evaluated how options for treating Hanford tank waste and for accelerator transmutation of wastes would fit with waste acceptance criteria for geological disposal. Dr. Thomas is a member of the American Chemical Society's division of nuclear chemistry and technology and the Network for Women in Science, and she has served on the DOE advisory committee on nuclear and radiochemistry education. In 2000, she received a LANL outstanding mentoring award for her work in fostering career development of women and members of her community. Dr. Thomas received her Ph.D. in nuclear chemistry as a student of Glenn Seaborg and her Master of Bioradiology, both from the University of California–Berkeley. She received her A.B. in chemistry from Middlebury College.

Appendix E

Acronym List

AEA	Atomic Energy Act (1954)
AEC	Atomic Energy Commission
ALARA	As Low As Is Reasonably Achievable
ANPR	Advance Notice of Proposed Rulemaking
ARARS	Applicable or Relevant and Appropriate Requirements
BEIR	Biological Effects of Ionizing Radiation
BRWM	Board on Radioactive Waste Management of the NRC
BSS	International Basic Safety Standards for Protection against Ionizing Radiation and for the Safety of Radiation Sources
CAA	Clean Air Act
CERCLA	Comprehensive Environmental Response, Compensation and Liability Act (1980, known as "Superfund")
CFR	Code of Federal Regulations
CID	Central Internet Database
CLSM	Controlled Low-Strength Material (used as a filler for waste disposal)
COWAM	Community Waste Management
CRCPD	Conference of Radiation Control Program Directors
CWA	Clean Water Act
DHEC	South Carolina Department of Health and Environmental Control

DOD	U.S. Department of Defense
DOE	U.S. Department of Energy
DOT	U.S. Department of Transportation
ECCS	Emergency Core Cooling System
EIS	Environmental Impact Statement
EM	Office of Environmental Management (DOE)
EPA	U.S. Environmental Protection Agency
ERDA	Energy Research and Development Administration
ERDF	Environmental Restoration Disposal Facility (at Hanford, Washington)
EU	European Union
EW	Exempt Waste
FFCF	Fuel Cycle Facilities Forum
FRC	Federal Radiation Council
FUSRAP	Formerly Utilized Sites Remedial Action Program
GAO	U.S. Government Accountability Office
GTCC	Greater-Than-Class-C
HLW	High-Level Waste
HPS	Health Physics Society
HSWA	Hazardous and Solid Waste Amendments
IAEA	International Atomic Energy Agency
ICRP	International Commission on Radiological Protection
ISCORS	Interagency Steering Committee on Radiation Standards
LAW	Low-Activity Waste
LILW—LL	Low and Intermediate Level Waste—Long Lived
LILW—SL	Low and Intermediate Level Waste—Short Lived
LLW	Low-Level Waste
LLWPA	Low-Level Waste Policy Act (1980, amended 1985)
MARSSIM	Multi-Agency Radiation Survey and Site Investigation Manual
MCL	Maximum Contaminant Level
MCLG	Maxiumum Containment Limit Goals
MIMS	Manifest Information Management System
MLLW	Mixed Low-Level Waste
MOU	Memorandum of Understanding

NARM	Naturally Occurring and Accelerator-Produced Radioactive Material
NCRP	National Council on Radiation Protection and Measurement
NEPA	National Environmental Policy Act (1969, amended 1970)
NESHAPs	National Emissions Standards for Hazardous Air Pollutants
NMA	National Mining Association
NNSA	U.S. National Nuclear Security Administration
NORM	Naturally Occurring Radioactive Materials
NPDES	National Pollutant Discharge Effluent Standards
NRC	National Research Council
NRSB	Nuclear and Radiation Studies Board of the NRC
NTS	Nevada Test Site
NWPA	Nuclear Waste Policy Act (1982)
OAS	Organization of [USNRC] Agreement States
OSRP	Off-Site Source Recovery Program
OSWER	Office of Solid Waste and Emergency Response
RPP	Radiation Protection Program
RCRA	Resource Conservation and Recovery Act (1976, amended 1984)
Rem	Roentgen Equivalent Man
ROD	Record of Decision
SARA	Superfund Amendments and Reauthorization Act of 1986
SECC	Southeast Compact Commission
SDWA	Safe Drinking Water Act
SKB	Swedish Nuclear Fuel and Waste Management Company
SNF	Spent Nuclear Fuel
Superfund	Hazardous Substance Response Trust Fund (CERCLA)
TENORM	Technologically Enhanced NORM
TRU	Transuranic
TSCA	Toxic Substances Control Act
UMTRCA	Uranium Mill Tailings Radiation Control Act
USACE	U.S. Army Corps of Engineers
USNRC	U.S. Nuclear Regulatory Commission
USRTR	U.S. Radiological Threat Reduction Initiative
WIPP	Waste Isolation Pilot Plant